T0001952

ORIGINS

ORIGINS

COSMOS, EARTH, AND MANKIND

HUBERT REEVES, JOEL DE ROSNAY,

YVES COPPENS, AND DOMINIQUE SIMONNET

FOREWORD BY JOSEPH SILK

First published in France under the title *La plus belle histoire du monde*

Helios Press books may be purchased in bulk at special discounts for sales promotion, corporate gifts, fund-raising, or educational purposes. Special editions can also be created to specifications. For details, contact the Special Sales Department, Arcade Publishing, 307 West 36th Street, 11th Floor, New York, NY 10018 or info@skyhorsepublishing.com.

Helios Press® is a registered trademark of Skyhorse Publishing, Inc.®, a Delaware corporation.

Visit our website at www.skyhorsepublishing.com.

10 9 8 7 6 5 4 3 2

Library of Congress Cataloging-in-Publication Data is available on file.

ISBN: 978-1-61145-507-6

Printed in China

CONTENTS

Foreword

Imagine a crisp, clear winter night in the countryside. I see the breathtaking vista of tens of thousands of stars, points of light glittering in the blackness. So many brilliant stars are visible that I can barely discern the constellations. I lie on my back and imagine that I am instantaneously transported far back in time. Would I see the same display of shining points of light? Or would it be even more spectacular? Would the stars shine even more brilliantly?

In my imagination, I view the stars not as I see them in the sky but as they were millions, even billions, of years ago. I can reach back to the beginnings of man and beyond, to the creation of the Earth and even of the Milky Way itself. Then the stars shone more intensely. The stars were youthful, vigorous, even brighter than those I see in the sky. And they were more crowded together, for space has been relentlessly flying apart as the universe expands.

But my imagination knows no bounds. Let me go back even further. What was there even earlier, before stars were born? The glory of the stars is a transient thing. Stars are born, shine brilliantly, exhaust their inner energy supply, and die. This we know, for we view the life and death of stars around us. But what, if anything, preceded the stars? What were the first objects in the universe? And what came next? Once stars were born, how did the conditions arise that led to the formation of planets, and ultimately of life? How did life evolve? Why did life evolve? Did life evolve painstakingly slowly over untold eons, or did life emerge in a sudden proliferation of species triggered by some cosmic event?

These are some of the issues that are discussed in this wonderful book. The interfaces of astronomy, physics, chemistry, paleontology, and anthropology, all crucial disciplines for probing the primordial fog, are explored in a series of penetrating responses to the queries of a leading science editor. So relax and enjoy this approach to some of the most fascinating topics imaginable.

Joseph Silk
Professor of Physics and Astronomy
University of California, Berkeley

ORIGINS

Prologue

Where do we come from? What are we? Where are we going? These are indeed the only questions worth asking. Each of us in our own way has sought the answer when we gaze upon a twinkling star, contemplate the incessant movement of the ocean tides, capture a woman's passing glance or the smile of a newborn babe. . . . Why are we alive? Why does the world exist? Why are we here?

Until recently, only religion, faith, belief have offered answers, provided a meaning. Today science, too, has come up with an opinion. That is doubtless one of the great accomplishments of our waning century: at long last we have a complete record of our origins. Science has managed to reconstitute the history of the world.

What, in fact, has science discovered that is so extraordinary? It has been able to determine that the same adventure has been going on for some fifteen billion years, an adventure that brings together the universe, life, and humankind, as if they were three chapters of one long epoch. The thrust of that same evolution, from the Big Bang to the emergence of intelligent life, has been a constant movement from the simple to the increasingly complex: the first particles, the atoms, the molecules, the stars, the cells, the organisms of living beings, until we arrive at the curious creatures we call human. . . . All stages follow one another in one continuous chain, all are swept along by the same movement. We descend from monkeys and bacteria, but also from the stars and the galaxies. The same elements of which our bodies are composed are those that, billions of years ago, brought the universe into being.

The idea is obviously upsetting, for it calls into question a number of ancient, long-held beliefs, as it strips away all sorts of prejudices. Since antiquity, the progress of knowledge has constantly taken man down a peg or two, put him in his rightful place. We thought we were at the center of the universe? Galileo, Copernicus, and others came along to disabuse us: actually, it turned out, we dwelled on a very ordinary planet on the outskirts of a relatively modest galaxy. We thought we were the products of original creation, quite apart from all other living species? Think again! Darwin set us firmly on the genealogical tree of animal evolution. Once again we were forced to swallow our misplaced pride and recognize that we are but the latest products of universal organization.

What this book intends is to describe, in easily understandable terms, this new history of the universe and the world (by which we mean the Earth), relying on the latest scientific knowledge. What we will discover as we go on in this tale is a surprising coherence. We shall see that the elements of matter will combine into ever more complex structures, which in turn will combine into more elaborate combinations, which in turn . . . This is the same phenomenon, the phenomenon of natural selection, that orchestrates each movement of this grand score, be it the organization of matter in the universe, the "game" of life on Earth, or even the formation of neurons in our human brains, as if there were a kind of "logic" to evolution.

Where does God fit into all this? A certain number of scientific discoveries conjoin with or reinforce intimate convictions. To be sure, we make every effort in the following pages not to mix science and religion, each of which reigns over separate domains. Science learns; religion teaches. Doubt is the motivating force behind the former, faith the glue that holds together the latter. Still, neither science nor religion is indifferent to the other. Our new history of the world makes no attempt to skirt, or avoid, either spiritual or metaphysical questions. Along our scientific journey we will perceive, at one point, a ray of biblical light, at another hear the echo of some ancient myth, and even, in the far distant African savanna, catch a glimpse of Adam and Eve. The latest discoveries, far from ending once and for all the debates between science and religion, bring them up to date, lend them new life and meaning. One can choose and conclude whatever one will.

Our book is based upon the latest scientific discoveries, which are the result of revolutionary tools: space probes exploring our solar system, the latest of which has sent back extraordinary new information about our sister planet Mars; spatial telescopes, the most famous of which is the orbiting Hubble telescope, which is increasingly ferreting out data relative to the secrets of the universe; the huge particle accelerators, which are in the process of retracing the earliest moments of the cosmos. And there are also our increasingly sophisticated computers, which have the ability to simulate the appearance of life, as well as various technological developments in the areas of biology, genetics, and chemistry, which have the ability to render the invisible visible, as well as reveal the infinitely small. Then, too, the recent discoveries of bone fossils, plus the vast improvement in dating techniques, have enabled us to reconstitute the progress of the various ancestors of the human race with astonishing precision. As recently as 1997, South African scientists brought to light the first footprints of an anatomically modern human, and even precious prehuman bones dating back 3.5 million years. And, no doubt, there is more coming.

Though our story is based on these latest findings, it is very much intended for the layperson, of whatever age or level of education. We have made every effort to avoid the mind-set or jargon of the specialist, as we have tried to eliminate any overly complicated terms. Nor have we refrained from asking questions, in the manner of an inquisitive child, that might be thought naive: How do we know about the Big Bang? How do we know what Cro-Magnon man ate? Why is the sky dark at night? Preferring not to take scientists at their word, we have constantly asked them to come up with the proof of whatever they say or maintain.

Each discipline is in search of an origin: the astrophysicists are trying to trace the origin of the universe; the biologists, the origin of life; and the paleontologists, the origin of the human race. Our story is cast as a play in three acts—the cosmos, life, and mankind—covering in time some fifteen billion years. Each act is divided into three scenes, and in each scene of this long adventure, as the play progresses chronologically, all the actors appear, be they inert or living. Each act is presided over by a leading scientist in the field.

In Act 1, our story begins. . . . But can we really speak of "beginning"? As the reader will see, this notion of a beginning is not incidental. On the contrary, it is the heart of the matter, the center of metaphysical debates, and poses the fascinating question of time. We'll approach the question at the farthest "point" back in time to which science can agree: fifteen billion years ago, the "time" of the Big Bang, that obscure light that preceded the stars. And we ask the question, the way a child might: what was there before the Big Bang?

From this "beginning," incandescent matter comes together through the action of astonishing forces that still preside over our destinies today. Where do these forces come from? Why are they immutable, whereas everything around them is in constant flux? Throughout our story they direct and control the grand universal machine. And, as the universe expands and cools off, these forces give rise to singular combinations—the stars and the galaxies—and at one point in the cosmic evolution, to a planet on the periphery of one such galaxy, which is destined to enjoy a very special success. What are these mysterious forces? Where does this irresistible movement from the simple to the complex come from? Did these forces exist *before* the universe?

The second act opens roughly four and a half billion years ago, on this very special planet located neither too close nor too far from a fortunate sun. Matter is going about its frenetic business of combining and recombining. On the surface of the

Earth, in new crucibles, another alchemy is beginning: molecules are joining together in structures capable of reproducing, giving birth to strange little droplets, then to the first cells, which come together into organisms that diversify, proliferate, colonize the planet, set in motion animal evolution, impose the force of life.

That life emanates from the inanimate is a difficult notion to accept. For centuries, the living world was considered too complex, too diverse, in short too "intelligent" to have appeared without the help of some kind of divine intervention. Today, the question has been settled once and for all: life, like inert matter, was part and parcel of the same process of evolution. Chance simply does not enter into it. How, then, did we move from the inert to life? How did evolution "invent" reproduction, sex, and death, the inseparable companion?

In the third act, in the lovely setting of the drying African savanna, the third avatar of the living world makes its appearance: the human species. Animal, vertebrate, mammal, primate . . . That we are all descended from African monkeys has long since been established. We are sons and daughters of monkeys, therefore, or rather of that ancient person who, several million years ago, first stood upright and began to view the world from a more lofty perch than its fellow creatures. But what made this prehuman ancestor stand up? What forces conspired to make him or her take that monumental step.

To be sure, for over a century we have known of our simian ancestry, which we have grappled with and tried to accept. But during the past several decades, the science of human origins has made such progress that the genealogical tree has been rudely shaken: a few hairy species have even fallen off. Today, we have the means at our disposal to pinpoint with great precision both the time and the place where the human comedy began. As if it had taken up the baton of matter, the prehuman and human species has needed no more than a handful of cosmic years—a mere few million—to evolve in turn and to invent all manner of things of increasing complexity: tools, hunting, war, science, art, love, and that strange propensity to constantly question both himself and the world around him. How did the human species make these various discoveries? Why did the prehuman and human brain constantly expand and develop? What happened to those human ancestors who failed to survive?

Our story, of course, is far from finished. One might even say it is just beginning. For it seems clear that the inexorable movement from the simple toward the complex is still going on, as is evolution itself. Thus we cannot conclude our story without asking one final question as we approach the millennium: where are we headed? In what direction, or directions, will this long adventure—which has been cosmic, chemical, biological, and now cultural—continue? What is the future of man, life, the universe? Science, of course, does not have all the answers, but it can make some interesting predictions. How will the human body go on evolving? What do we know about the future evolution of the universe? Are there other forms of life out there, somewhere in the vast universe? These are questions all four of us will discuss and ponder in the Epilogue.

One final cautionary note: we have made every effort throughout this book not to yield to the determinist temptation. And we ask the reader's indulgence if, upon occasion, in an effort to make some points clearer, we have resorted to terms that may seem shocking or out of place: no, we cannot say that matter "invents," that nature "manufactures," or that the universe "knows." That "logic" of organization is simply stating a fact. Science specifically refrains from imputing any intentionality to its findings. Let each reader make up his or her mind on that score. If our story seems, despite all we have just said, to have a meaning, we nonetheless cannot go so far as to maintain, or conclude, that our appearance was ineluctable, at least on this planet. Who can say how many unsuccessful paths evolution has taken before celebrating our birth? And who can deny the extreme fragility of that result, the human species?

We think of ourselves as the end product of this long evolutionary process because the story we tell is ours. We bear the universe in the very depths of our being: our human bodies are composed of atoms of the universe, our cells enclose a parcel of the primitive ocean, our genes are for the most part identical to those of our fellow primates, our brains contain the various strata of the evolution of intelligence, and when, in the mother's womb, a child is conceived, it passes through the full stages of animal evolution, albeit in a speeded-up form.

But no matter how we view our origins, whether our vision is mystical or scientific, determinist or skeptical, religious or agnostic, there is one undeniable fact that we must constantly strive to bear in mind, one overriding moral consideration: we are but a flickering spark in the overall context of the universe. May we have the wisdom not to forget that.

Dominique Simonnet

Act 1

The Universe

Scene 1

Chaos

The stage is white, infinite. Everywhere, there is nothing but an implacable clarity, the light of an incandescent universe, the chaos of a matter that as yet has neither meaning nor name. . . .

But What Was There "Before"?

DS: An explosion of light back in the furthest reaches of time is where our story begins, the origin of the universe, which science has been focusing on and speculating about over the past several years. Before we can consider that phenomenon, however, we have to stop and ask ourselves this naive question: what was there before?

HR: When you bring up the subject of the beginning of the universe, you inevitably come up against a problem of vocabulary.

For us, the word "origin" relates to an event that can be situated in time. Our personal "origin," for example, is the moment when our parents conceived us. That origin has both a "before" and an "after." We can date it, note it down specifically in the context of our personal story. And we are willing to accept the fact that the world existed before we came into it.

But here we're talking about the origin of origins, the very first . . .

And therein lies the great difference. The origin to which we're referring, the beginning of the beginning, cannot be thought of as an event comparable to any other. We find ourselves in the same situation as the early Christians, who kept asking what God was doing before He created the world. The popular response in those days was: "He was busy preparing hell for all the people who ask that question!" Saint Augustine did not agree. He clearly saw the inherent difficulty in such a question, which presupposed that time existed *before* the Creation. His answer was that Creation was not only the beginning of matter but also the beginning of time. That point of view is very close to what science is saying today. Space, matter, energy, and time are all inextricably intertwined, indissoluble. In our cosmologies, they appear together. If there is an origin of the universe, then that is also when time began. Therefore, there is no "before."

"*If* there is an origin of the universe," you say. Which implies that there's some doubt about it.

The major discovery of the twentieth century is that the universe is neither immutable nor eternal, as most scientists believed in the past. Today we are convinced of that notion: the universe has a history, it has constantly, endlessly evolved, become rarefied, grown cooler, become more structured. Both our observations and our theories allow us to go back in time and reconstruct the story of how the universe has evolved. Those observations and theories confirm that this evolution has been going on for a very, very long time: somewhere between fourteen and fifteen billion years, according to the best estimates. We now have at our disposal a sufficient number of scientific elements to describe what the universe was like at that time: it was completely disorganized; there were no galaxies or stars, no molecules or atoms or even the nuclei of atoms. It was nothing more than a kind of thick puree, a formless, pasty soup, with temperatures in the billions and billions of degrees.

And nothing before?

We don't have any knowledge of what preceded that event, not even the faintest clue that would enable us to delve deeper into the past. All the observations, all the data gathered by astrophysicists stop at that same frontier. Does that mean that the universe "began" fifteen billion years ago? Does it mean that the Big Bang is really the origin of the origin? We have no idea.

And yet that is what students are being taught in school today: the universe began with a Big Bang roughly fifteen billion years ago. And that, in fact, is what scientists have been telling us over and over again for many years now.

We probably didn't express ourselves clearly enough, and we've been misunderstood. We could speak about a beginning, a veritable first moment, if we were sure that there was nothing before. The fact is, at those high temperatures, our notions of time, space, energy, of temperature itself, no longer apply. Our laws no longer function; we are completely stripped bare.

When you say that, aren't you—scientists in general—begging the question, copping out? When we tell a story, there's always a beginning. Since our subject here is the "story" of the universe, it's not all that unreasonable for us to go looking for the point when it all began.

Of course, in the human context, all stories do have a beginning. But we have to beware of extrapolations. We can say the same thing about Voltaire's clock: the very existence of the clock, he maintained, proved the existence of a clock maker. Does this reasoning, however unassailable at our level, the human scale, really apply to the "clock" of the universe? I'm far from sure that it does. What we have to consider is whether or not our logic, as Heidegger said, is the supreme instance, if the assertions and arguments that are valid here on Earth can be fairly applied to the universe in its entirety. The only real question is that of our existence, that of reality, of our consciousness: "Why is there something rather than nothing?" Leibniz asked. But that is a purely philosophical question, which science is incapable of answering.

The Horizon of Our Knowledge

To get around this brainteaser, could we therefore define the Big Bang as the beginning of space and time?

Let's rather define it as the moment when these notions became usable. In reality, the Big Bang is our horizon in time and in space. If we assume that that is the point zero of our story, it's for the sake of convenience; it's because we have nothing better to go on. We're like the early explorers facing the vast ocean: we can't see whether there's anything beyond the distant horizon.

If I understand correctly, the Big Bang is in fact a manner of designating not actually the limit of the world but the limit of our knowledge.

Precisely. But be careful: having said that, we cannot therefore go on to conclude that the universe does not have an origin. Once again, we simply don't know. But for the sake of argument, and to simplify the question, let us assume that our adventure began fifteen billion years ago, in this infinite and unformed chaos that will slowly structure itself. That, in any case, is the beginning of our story of the world such as science can reconstitute it today.

Specialists can make do with an abstraction to portray the Big Bang. But we laymen need a metaphor. We've often heard it described as a ball of concentrated matter that exploded in an enormous burst of light and filled the entire space.

Because some scientists describe it that way does not mean it's right. That explanation would presuppose the existence of two spaces: one filled with matter and light that progressively invade a second space, which is empty and cold. In the model of the Big Bang, only one space is produced, uniformly filled with light and matter that are expanding in every direction: all its points are moving away from one another at the same speed.

It's difficult to conceive. Is there some kind of visual description you can give the Big Bang?

You can, in a pinch, imagine an enormous explosion, but only if you can accept the notion that the explosion is occurring in each and every point of a vast and perhaps (but not certainly) infinite space. Of course, that's extremely difficult to imagine, but is there anything so surprising about it? When we grapple with such matters on such a scale, our intellectual powers find themselves in unusual, uncharted territories, and our descriptions are of necessity inadequate at best.

And God?

Infinite or not, that image corresponds very nicely to the Bible's description of the creation of the world. "And there was light. . . ."

That striking similarity actually worked against the acceptance of the Big Bang theory when it was first proposed in the early 1930s, especially after the statements made by Pope Pius XII that science had rediscovered *Fiat Lux*, "Let there be light," and thus validated the Bible's description of the Creation. In Moscow at the same time, the Communists' attitude was also revealing. At first they completely dismissed the pope's pronouncements as "papal stupidities"; then they realized that this idea could serve to validate and confirm the Communist dogma of "historical materialism." So there was a rapid shift in the party line, and Lenin was given credit for having foreseen the discovery. Nonetheless, despite all these attempts by both religious and political groups to co-opt the idea for their own ends, the theory prevailed. In the course of the following decades, proof after proof piled up, and virtually all astrophysicists came to accept the theory as the best explanation of the history of the cosmos. One exception is the English astrophysicist Fred Hoyle; he continues to maintain that the universe is stationary. Ironically, it was he who derisively baptized the theory the "Big Bang," and the term stuck.

*

That science meets religion along its way is not so terrible, is it?

So long as one makes sure not to confuse their different approaches. Science seeks to understand the world; religions generally view their mission as trying to give life a meaning. They can each shed light on the other, but only so long as they both remain in their own territory. Each time the church has tried to impose its explanation on the world, there has been a conflict. Remember what Galileo said to the theologians who set themselves up against him: "You can tell us how to gain entrance into heaven. Let us scientists tell you how things are faring in the heavens above." Remember, too, the staunch opposition of the church fathers to Darwin's theories. Science is concerned with measurable, deducible facts. It does not attempt to deal with what might exist beyond the visible world, the measurable. Contrary to popular opinion, science does not set itself up in opposition to God. Science can neither prove nor disprove His existence. That subject transcends the limits of science.

Despite all you say, not only Christianity but also several other religions and mythologies explain the creation of the world by an explosion of light. Don't you find that disturbing?

The image of an initial chaos that progressively transformed itself into an organized universe does indeed occur in a number of traditional versions of how the world began. We find it in the writings of the Egyptians, in the oral traditions of the North American Indians, among the stories of the ancient Sumerians.

This chaos is often depicted by an aquatic image, for instance, an ocean in darkness. "Nothing existed except the empty sky above and the calm sea below, in the depths of night," according to the Mayan tradition. "And the entire Earth was nothing but sea," says a Babylonian text. And Genesis: "And the Earth was without form, and void; and darkness was upon the face of the deep. And the spirit of God moved upon the face of the water." The metaphor of the egg was also used. Inside the egg, a seemingly formless liquid is transformed into a baby chick. It's a lovely image of the evolution of the universe. Among the Chinese, the egg separates into two halves, one of which becomes the heavens, the other Earth. Still, in all these mythological descriptions, chaos is associated with water and darkness. In modern cosmology, chaos is made of heat and light.

And yet these analogies between the scientific description and early myths are undeniable.

Could it be a matter of pure coincidence? Or the result of some intuitive knowledge? After all, as we shall see in the course of our story, we ourselves are made up of the dust of the Big Bang. Could it be that we bear within ourselves the memory of the universe?

The Discovery of History

How did science happen to propound the idea of an original chaos and the evolution of the universe?

For two thousand years, philosophical tradition had it that the universe was eternal and unchanging. Aristotle expressed himself very clearly on the subject, and his ideas dominated Western thought for more than two thousand years. According to him, the stars were made of unperishable matter, and the heavenly map was fixed and unchanging. Today, thanks to modern instruments, we know that Aristotle was wrong. Stars are born and die, after having lived for millions or even billions of years. They shine, consuming their nuclear fuel, and when it is used up, they go out, disappear. We can even date them.

No one ever came up with the idea, early in human history, that the heavens could change?

Actually, some did. Several philosophers suggested as much, but their views never took hold. Lucretius, the Roman philosopher who lived in the first century before Christ, asserted that the universe was still in its youth. How did he arrive at that notion, which was so far ahead of its time? He followed a very clever line of reasoning. Since I was a child, he said to himself, I've noticed all around me that improvements have been made in various areas. We've improved the sails of our ships, so that they sail faster; we've invented weapons that are more efficient; we've invented musical instruments that are more and more sophisticated. If the universe were eternal, all these changes and improvements would have had time to occur a hundred times over, a thousand times, millions of times! In which case I would be living in a finished world, which no longer changed. But the fact is, during the brief years of my own life I've witnessed so many improvements that it proves that the world has not always existed.

A pretty solid piece of deduction.

Today, cosmology confirms Lucretius's notion on three grounds: (1) the world has not always existed, (2) it is constantly changing, and (3) this change is expressed by the movement from less efficient to more efficient, that is, from the simple to the complex.

Going Back in the Time Machine

Modern science is founded on what discoveries?

Thanks to our instruments—instruments of both physics and astrophysics—we have been able to discover traces of the universe's past. We can reconstitute its history in the same way anthropologists can reconstitute humankind's past from fossils found in sedimentary layers that later become exposed through tectonic shifts, weathering, and so forth. But we have an enormous advantage over students of prehistory: we can actually *see* the past.

Could you clarify that?

In our scale, light travels very fast: the speed of light, as we all know, is 300,000 kilometers—186,000 miles—per second. On the scale of the universe, this speed is very low. Light comes to us from the Moon in roughly a second. Light from the Sun reaches us in eight minutes. But from the closest star it takes four years; light from Vega takes eight years, and from certain galaxies it takes billions of years. Modern-day telescopes enable us to observe very distant stars, quasars, for example, the luminosity of which can be as much as ten thousand times that of our entire galaxy. Some of those quasars are located twelve billion light-years away from us. In other words, we're seeing them today the way they were twelve billion years ago.

Whenever you focus your telescope on any given region of the universe, what you're really doing is observing a moment of its history.

That's exactly right. The telescope is an instrument that allows us to go back in time. In contrast to historians, who can never actually see Rome the way it was in its glory, we astrophysicists can truly *see* the past and observe stars the way they were. We see the Orion Nebula the way it was at the end of the Roman Empire. And the galaxy of Andromeda, visible by the naked eye, is an image that's two million years old. If the people who dwell in the Andromeda galaxy were looking at planet Earth today, what they'd be seeing would be the time of prehistoric man.

All of which means that the sky we gaze upon at night, the stars we see, the myriad stars and galaxies, are really only so many illusions, the superimposition of past images.

Strictly speaking, we can never see the present state of the world. When I look at you, I'm seeing you as you were a hundredth of a microsecond ago, the time it takes for light to reach me. A hundredth of a microsecond is very long on the atomic scale, even though it's imperceptible to our minds. But human beings do not disappear in that lapse of time, so I can hypothesize without risk that you are still there. The same goes for the Sun: it does not change during the eight minutes it takes its light to reach Earth. The stars we see at night with the naked eye, the stars in our own galaxy, are also relatively close. But for the distant stars, those that we can detect only with the most powerful telescopes, it's a whole other matter. The quasar that I see at twelve billion light-years' distance in all likelihood does not exist today.

Would it be possible, therefore, to go even further back in time, to that famous "horizon" when the Big Bang took place?

*

The further back we go into the past, the more opaque the universe becomes. Beyond a certain limit, light can no longer reach us. That horizon corresponds to a time when the temperature was roughly three thousand degrees Celsius. According to the conventional time clock based on the Big Bang, the universe would then already be about 300,000 years old.

Proofs of the Big Bang

In other words, the Big Bang remains extremely abstract. One can even ask if it isn't a pure product of the scientific imagination, if there is any reality behind it.

Like all scientific theories, the theory of the Big Bang is based both on various observations and on a mathematical system (Einstein's general theory of relativity) capable of reproducing the numerical values. If this theory is credible, it's because it has already correctly predicted the result of several observations, and because these predictions have now been confirmed, which shows not only that the Big Bang is a product of scientific imagination but also that it touches on the reality of the world.

Let's assume that what you say is true. But how can you describe the Big Bang if you can't see it?

We do see a number of manifestations that emanate from it. In 1929, American astronomer Edwin Hubble established that the galaxies were moving away from one another at speeds proportional to the distances between them—a little like a raisin cake that you put in the oven: as it rises, the raisins move away from one another. This movement of all the galaxies, which we have labeled the expansion of the universe, has been verified up to speeds of tens of thousands of miles per second. According to Einstein's theory of relativity, this expansion translates into a progressive cooling of the universe. Its actual temperature is three degrees absolute, that is, minus 270 degrees Celsius. And this cooling process has been going on for some fifteen billion years.

How do we know?

Let's try to reconstitute the scenario by turning the film backward. The further we go back in time, the closer together the galaxies are: the universe becomes denser and denser, hotter and hotter, more and more luminescent, until we reach that point, about fifteen billion years ago, where the temperature and the density attain gigantic numbers. That is what is conventionally called the Big Bang.

The point at which our cake is a ball of dough . . .

Beware of comparisons; they can be deceiving. The analogy of the raisin cake implies that the universe was smaller than it is today. Nothing is less certain. It could well be infinite and always have been infinite.

*

Hold on! How can you conceive of a universe that was infinite originally and from that point on grew larger?

The word "large" is meaningless when you're talking about infinite space. Let us simply say that it is going to become more and more rarefied. To try to make the idea clearer, we can conceive of a universe in a single dimension: a calibrated ruler that extends to infinity both left and right. Now let's imagine that the ruler begins to expand, that is, that each mark on the ruler begins to move away from its neighbor. The points on that ruler are becoming farther and farther apart from one another, but the ruler itself remains infinite.

I can imagine that the discovery of the galaxies' movement away from one another is not the only proof of the Big Bang.

There are, in fact, several others. Take, for instance, the age of the universe. We can measure it in several ways: by the movement of the galaxies, by the age of the stars (by analyzing their light), and by the age of atoms (by calculating the proportion of certain atoms among them that disintegrate with the passage of time). The idea of the Big Bang requires that the universe be older than the oldest stars and the oldest atoms. And in all three cases, when we make our measurements, we come up with a number in the vicinity of fifteen billion years, which reinforces the credibility of our theory. The latest measurements of the expansion of the universe by the space satellite *Hubble* (1997), together with the reevaluation of the age of the oldest stars by the satellite *Hipparchos* (1997), have confirmed this point of view. Besides which, we also have our "fossils."

The Fossils of Space

Fossils? I assume it's safe to say you're not referring to shells or bones.

No, what we call fossils are physical phenomena that date from the earliest times of the cosmos, the characteristics of which allow us to reconstitute the past, the way paleontologists do with their bone fragments. Take but one example: the "background radiation," also called the "fossil radiation," which was emitted at a time when the temperature of the universe was several thousand degrees Celsius. That's a vestige of the extraordinary light that existed in the Big Bang, a pale light evenly distributed throughout the universe. That radiation reaches us in the form of radiomillimetric waves that can be picked up by the proper antennae pointing in every direction toward the heavens. That's the image of the cosmos fifteen billion years ago, the oldest image in the world.

What you're saying, then, is that the space between the stars is not empty?

Light is made up of particles we call "photons." Each cubic centimeter of space contains roughly four hundred of these grains of light, most of which have been traveling since the very earliest periods of the universe, the others having been emitted by the stars.

How do you count these photons?

Actually, we measure the temperature of this radiation. We can do that with a high degree of accuracy, thanks especially to deep-space radio-wave probes. It is 2.716 degrees absolute. There is a simple relation between the temperature and the number of photons, and the calculation gives us 403 grains of light in every cubic centimeter of space. The intensity of the flux confirms this value. Neat, no?

Impressive, I must say.

I should add that the existence of this fossil radiation had been predicted by astrophysicist George Gamow in 1948, seventeen years before it was actually observed. This radiation was, according to Gamow, a necessary consequence of the Big Bang theory.

What that theory predicted is therefore consistent with what we observe today?

The Hubble telescope out in space has provided us with a number of confirmations. For example, thanks to the Hubble telescope, we can now see a distant galaxy the way it was when the universe was warmer, and we are able to determine the temperature of the radiation in which a galaxy twelve billion light-years away is steeped. That temperature is 7.6 degrees absolute—which is precisely what the theory predicted. During the time it took for the light from that galaxy to reach us, the temperature fell to 2.7 degrees, which proves that we are living in a universe that is in the process of cooling off.

The Darkness of Night

Other examples?

Here's one: helium atoms are also fossils; their populations relative to hydrogen in the universe are also consistent with the theory, indicating that the universe of the past has reached temperatures of at least ten billion degrees Celsius. There are also indirect proofs, such as the darkness of the sky at night.

How is that a proof of the evolution of the universe?

If the stars were eternal and unchanging, as Aristotle maintained, the quantity of light that they would have emitted during an infinite period of time would also be infinite, in which case the sky should be extremely luminous. Why isn't it? This enigma puzzled and tormented astronomers for centuries. We now know that if our sky is dark, it's because the stars did not always exist. Fifteen billion years is not a long enough period to fill the universe with light, especially when the space between stars is constantly increasing. The darkness of night is a supplementary proof of the evolution of the universe.

Any more?

Another indirect argument in favor of a universe undergoing constant change comes to us directly from the theory of relativity. This theory, formulated by Einstein in 1915, does not allow the universe to be static. If Einstein had been able to interpret correctly the message of his own equations, he could have predicted that our universe was in evolution fifteen years before others made that discovery.

In short, today, nobody opposes the theory of the Big Bang?

Let's say that among the various possible explanations of the origin of the universe, the Big Bang is far and away the best choice. No other rival scenario explains as simply and as cogently the impressive array of observations that have been made. No other theory has made as many predictions that have been borne out. To be sure, the scenario of the Big Bang is far from being completely satisfying; there are still many weak points and unanswered questions. One should think of it as a concept still very much in progress, still being perfected, filled with all sorts of hesitations. We are still feeling our way, and in the future we'll doubtless modify the theory in various respects, perhaps incorporate it into a vaster scheme of thought. But the basis, the main points, should remain intact.

*

When you say "main points," what exactly are you referring to?

To a few simple affirmations: the universe is not static; it is growing colder and becoming more and more rarefied. But above all—and for us this is a crucial element—matter is becoming progressively more organized. Particles dating from the earliest times of the universe are coming together to form structures that are more and more elaborate. Just as Lucretius predicted, we are moving from the simple to the complex, from the less efficient to the more efficient. The story of the universe is the story of matter becoming more and more organized.

Scene 2

The Universe Gets Organized

In order of appearance on stage: minute particles, in an indescribable state of disorder; then, as a result of their coupling, the first atoms, which also attempt explosive liaisons in the heart of burning stars.

Alphabet Soup

The story of complexity begins. We are at the horizon of our past, some fifteen billion years ago. What is the universe made up of at this moment?

The universe is a homogenous puree of elementary particles: electrons (those that provide electrical current), photons (the grains of light), quarks, neutrinos, and a panoply of other elements called gravitons, gluons, and so forth. We call these "elementary," because we cannot break them down into smaller elements, at least so far.

The usual term to describe this, I believe, is primitive puree, which means that the whole thing is a mixture, disordered and disorganized.

I like to compare it with the alphabet soup I used to eat as a child, when the game was to try to write names with the letters floating in the broth. In the universe, these letters—that is, the elementary particles—are going to come together to form words, which in turn will join to make sentences, which will later become paragraphs, chapters, books. At each level, the elements will regroup and form new structures at a higher level. And each new structure will have properties that its individual elements did not themselves possess. We refer to that as "emerging properties." Quarks assemble themselves into protons and neutrons, which later will come together to form atoms, which will form simple molecules, which will form more complex molecules, and so on. That's the alphabet pyramid of nature. But that all happened later.

How long did this whole process take?

During the first dozens of microseconds following the Big Bang, the universe was a vast magma of quarks and gluons. At roughly the fortieth microsecond, when the temperature descended below 10^{12} degrees (a million million degrees), the quarks came together to form the first nucleons: the protons and neutrons.

The First Second

What extraordinary precision! How in the world can we know about the first second of the universe, not to mention the smallest fractions of that first second, when we're not even sure whether the universe is fourteen or fifteen billion years old?

You have to understand the precise meaning of the words. The "first second" indicates the period when the temperature of the universe was ten billion degrees. Before the first second, the temperature was even higher. The difficulty is situating that precise second in our history: let's say, for argument's sake, fifteen billion years ago. The big particle accelerators allow us to reconstitute, for a few brief moments, the great densities of energy that existed at that time. They correspond to temperatures of 10^{16} degrees. In the cosmic scenario, these densities obtained for only a micro-microsecond. But once again, we're talking about a chronology that is meaningful only in the context of the Big Bang theory. It's a conventional clock, a kind of landmark.

*

And yet we noted that physics reaches its limits when it comes to the moment of the Big Bang; it simply has no ready answers.

In physics we have two solid theories: quantum physics, extremely precise, which describes particles, provided they are not subjected to too strong a field of gravity; and Einstein's theory of gravity, which takes into account the movement of the stars but ignores the quantum behavior of particles. The limits of physics are calculated to be at temperatures of approximately 10^{32} degrees (Planck's temperature). At that temperature, particles are in fact subjected to very powerful fields of gravity. Calculating their properties has so far been beyond us. In the past fifty years of trying, no one has been able to solve the problem. What we need is another Einstein.

Meanwhile, until another Einstein comes along, let's focus on that first cosmic second. Why didn't the universe remain in a state of puree? What made it start evolving?

It was the four forces of physics that presided over the agglomeration of particles, then of atoms, molecules, and the great celestial structures. The nuclear force, also called the strong force, unites the atomic nuclei; the electromagnetic force causes the cohesion of the atoms; the force of gravity organizes the grand-scale movements—those of the stars and the galaxies; and the weak force produces certain kinds of radioactive decay and intervenes at the level of the particles known as neutrinos. But in the very first moments of the universe, the temperatures were so high as to keep anything from joining together to form structures, the same way that our normal summer temperatures on Earth keep water from turning into ice. The universe therefore had to cool off before the four forces could move into action and make the initial attempts at combining matter.

The Force Is with Us

But where do these four forces come from?

That's a vast question, one that gets us very close to metaphysics. Why do these forces exist? Why do they have the mathematical properties that we have discovered? We do know today that these forces are the same everywhere, here on Earth and at the outer limits of the universe, and that they have not changed one iota since the Big Bang. Which raises a big question mark in the context of a universe where everything is in flux, everything is changing.

How do we know that these forces are unchanging?

We've been able to verify it in several ways. Several years ago in Gabon, mining engineers discovered a uranium deposit, the composition of which was very special. All evidence indicated that this mineral had been subjected to intense neutron radiation. About a billion and a half years ago, a kind of natural nuclear reaction had spontaneously occurred in this African mine. When we compared the amount of radioactivity in these atomic nuclei with the amount in the atomic nuclei of our reactors, we were able to show that the nuclear force at that time had exactly the same characteristics as it does today. In the same way, we can determine whether the electromagnetic force has changed by comparing the properties of the old and new photons.

How can you do that?

Spectroscopes allow us to detect the photons emitted by iron atoms coming to us from a distant galaxy. They are “old” photons that have been traveling for, let’s say, twelve billion years.

I find that notion difficult to understand. You mean to say that you can really receive old particles that you can “capture”?

Absolutely. In our laboratories, we can compare their properties with those of young photons emitted by an electric arc with iron electrodes. The result: the electromagnetic force has not changed during the twelve-billion-year period that separates these two generations of particles. In the same way, the analysis of the abundance of light nuclei shows that the force of gravity and the weak force have undergone not the slightest modification since the temperature of the universe was ten billion degrees, that is, fifteen billion years ago.

*

How can you explain that these forces are completely immutable?

Do you mean on what stone tablets, like those that Moses brought down from the mountain, are these laws written? Are they located "above" the universe, in that world of ideas dear to the followers of Plato? These are not new questions: people have been debating them for at least twenty-five hundred years. The progress of astrophysics has enabled us to bring the questions to the forefront of contemporary thinkers but has not enabled us to provide satisfactory answers. All we can say is that, contrary to the universe, which is in constant flux, these laws of physics do not change, either in space or in time. In the context of the Big Bang theory, these laws have presided over the making of complexity. What is more, the properties of these laws are even more astonishing. Their algebraic formulas and their numerical values appear particularly well adjusted.

In what way are they "well adjusted"?

Our mathematical simulations prove it: if they had been even slightly different, the universe would never have emerged from its initial chaos. No complex structure would have emerged, not even a molecule of sugar.

Why?

Just imagine for a moment that the nuclear force had been slightly stronger. All the protons would have quickly gathered into heavy nuclei. There would have been no hydrogen left to assure the longevity of the Sun and to form water masses here on Earth. Instead, the nuclear force was just intense enough to produce some heavy nuclei (including carbon and oxygen) but not so intense as to completely eliminate the hydrogen—just the right dose. One could even go so far as to say that complexity, life, and consciousness were implicit from the very first instants of the universe, that they were inscribed, so to speak, in the very form of the laws—not as being "necessary," but as being a possibility.

Isn't that a kind of a posteriori reasoning? We note today that these laws led to the evolution of humankind. That doesn't mean, or even imply, that these laws were made with that end in mind, does it?

That is indeed the sixty-four-million-dollar question: is there an "intention" implicit in nature? That's really not a scientific question but one that belongs to the realm of philosophy or religion. If you're asking me personally, my answer would be yes. But the question remains: what form does that intention take, and what *is* that intention? These are questions that interest me passionately. But the fact is, I have no answers. Allegorically speaking, one can say, if "nature" (or the universe, or reality) had had the "intention" of engendering conscious beings, it would have "done" precisely what it has done. Of course, that's a posteriori reasoning, but that doesn't make it any less interesting.

The Lesson of the Moon

How long have we known of the existence of the laws of nature?

It took us several centuries to recognize their existence. The Greek philosophers were already looking for the primordial elements that, according to them, presided over the making of the cosmos. Aristotle divided the world into two: the world "beneath the Moon" (our world), which was subject to change, where wood rotted and metal rusted; and the space "beyond the Moon," inhabited by the celestial bodies that were perfect, immutable, and eternal.

Everything was for the best in the best of all possible worlds.

That notion of the perfection of the celestial bodies influenced Western thinking for a very long time. The sunspots visible to the naked eye, of which the Chinese of antiquity were well aware, were never mentioned in the West until the time of Galileo. The sentence "I'll believe it when I see it" can also be reversed: "I'll see it when I believe it." When Galileo focused his telescope on the Moon and for the first time observed that there were mountains on the Moon, everything had to be called into question again. "The Moon is like the Earth. The Earth is a star. There are not two worlds, but one, and the laws that govern it are one and the same." Newton went even further: for him, it's the same force that makes an apple fall to the ground and keeps the Moon in orbit around the Earth and the Earth in orbit around the Sun. It's the "universal" force of gravity, which he would use to explain the movement of the planets. The laws of terrestrial physics operate throughout the universe.

But that accounts for only one force.

In the nineteenth century, scientists were already aware of the electrical force that attracted feathers to amber, as they were aware of the magnetic force that governed the needles of a compass. The work of a number of scientists proved that these seemingly different forces were in fact one and the same: the electromagnetic force that simply manifests itself differently under different circumstances. Over the course of the twentieth century, we discovered two new forces: the nuclear, or strong, force and the weak force. In 1970 or thereabouts, we proved that the weak force and the electromagnetic force were actually manifestations of the force called "electroweak." Physicists would love to unite all the various forces into one, but for the moment, that remains a dream.

We've discovered two forces in our century. Why shouldn't there be others waiting to be discovered?

It's entirely possible. Physicists keep an inventory of forces the way botanists index flowers. There's no reason to believe that we've completed our inventory. Ten years ago, the idea of a fifth force was theorized, but it failed to stand up to analysis.

The First Minutes

How did these four universal forces interreact at the beginning of our history?

When the temperature is extraordinarily high, thermal activity rapidly disassociates all the structures that can be formed. As the temperature drops, the forces enter into play according to their relative strength: first the nuclear force; quarks come together three by three to form nucleons (neutrons and protons), the particles that make up the nuclei of all atoms. This happened when the universe was about twenty microseconds old.

Why three by three?

These particles come together by chance, but certain combinations fail to hold. If they come together two by two, the couples they form are unstable and disintegrate rapidly. Only two kinds of trios resist: an assemblage of two "up" type quarks together with one "down" type, which form a proton; and two "down" type quarks with one "up" type, which form a neutron. The first trio resulted in hydrogen, which has a nucleus of only one proton. A bit later, the nuclear force incited the assemblage of two protons and two neutrons, to make up a helium nucleus. At this point, the temperature had dropped to ten billion degrees, and the universe was already one minute old.

It took only one minute for the first atomic nucleus to form!

The various forces can come into play only under certain conditions of temperature, a little like water turning into ice. If it's too hot, the forces don't act. After these first few minutes, the universe grew colder, again inhibiting the activity of the nuclear force. At this point, the composition of the universe was 75 percent hydrogen nuclei (protons) and 25 percent helium nuclei. As for the organization of the universe, nothing would happen now for several hundreds of millions of years.

One minute of activity and hundreds of millions of years of waiting. Talk about fits and starts!

Complexity does not move forward at a regular pace. When the temperature drops below three thousand degrees, the electromagnetic force enters into play. It sends electrons into orbit around the nuclei and thus creates the first atoms of hydrogen and helium. The disappearance of free electrons has the effect of making the universe transparent: the photons, these grains of light, are no longer affected by the matter of the cosmos. They wander in space and are progressively degraded into energy.

They are still there today, old and degraded, constituting the fossil radiation. Evolution made a second pause, and it was several hundred million years before it took off again.

The First Galaxies

What gave it the impetus this time?

Under the force of gravity, matter, which until now had been almost homogeneous, began to form clots. From the time electrons were picked up by the nuclei, the field was open, and grand-scale structures could be formed. Until then, any attempt to concentrate matter was quickly neutralized by the action of the radiation on the electrons. This time, it would be able to condense into galaxies.

Once again, I can't help asking: why?

I have to admit that we know relatively little about this period of the universe's history. In fact, astrophysicists refer to this period as the "dark age of cosmology." Observations made by the COBE satellite have shown us that, at this point in cosmological time, matter was not perfectly homogeneous and isothermal. Regions that were slightly denser than the mean played the role of "germs" of the galaxies. Their attraction progressively pulled surrounding matter toward them. There was thus a kind of snowball effect, and these denser regions became larger and larger, until they formed the magnificent galaxies that we see in today's night skies.

Did this phenomenon occur everywhere at the same time? Is there no desert anywhere in the universe?

In the universe there is a hierarchy of massive objects, beginning with clusters of galaxies, galaxies themselves, constellations of stars, and individual stars. Our solar system, for example, belongs to a galaxy we commonly refer to as the Milky Way, which is made up of billions of stars and, taken together, forms a disk 100,000 light-years in diameter.

A speck of dust in the universe . . .

Our galaxy belongs to a small cluster of a few tens of members (including Andromeda and the two Magellanic Clouds), which itself is part of a larger cluster of several thousand galaxies called the Virgo Cluster. In its center, this superagglomeration is home to a giant galaxy, a hundred times larger than our own, toward which other galaxies are drawn. We refer to that as a cannibal galaxy.

Charming thought . . .

On a scale larger than a billion light-years, the universe seems extremely homogeneous. Everything is more or less uniformly populated; there is no "desert," and nothing looks more like one part of the universe than any other part of the universe.

At that time, the face of the universe was therefore changing.

About a hundred million years after the Big Bang, the universe no longer looked like the homogeneous puree that it was in its earliest stages. Its physiognomy was very much like that we're familiar with: a vast space, not very dense, sprinkled with these superb galactic islands a million times denser than the rest of it. Within these galactic islands, matter was condensing through the action of the force of gravity and forming the stars. That caused the temperature to rise. Thus the stars avoided the general cooling that was going on all around them. They warmed up, emitting energy: the stars began to shine. The largest, fifty times more massive than our own Sun, would use up their atomic fuel in three or four million years. The less massive would go on living for billions of years.

Why did stars become round?

What does the force of gravity do? It attracts matter. What is the configuration in which all the elements are closest to one another? A sphere. That's the reason stars are spherical, as are the planets, except for those that are very small. Within any celestial body whose diameter is more than two hundred kilometers, the force of gravity dominates the chemical forces that give matter its rigidity and forces it to become spherical: the

Moon is round, as are the satellites of Jupiter. By contrast, the satellites of Mars, which are much smaller, do not have enough gravity to make their rocky masses round, which is why they are not spherical.

But galaxies aren't spherical. Why?

Their rotation flattens them out and gives them the disk shape with which we're familiar. Because of its rotation, Earth is slightly flattened, as is our Sun.

Why the Stars Don't Fall

Why have all these stars not been attracted to one another?

Newton asked himself that very question. Since stars are massive objects, he reasoned, they are mutually drawn toward one another. Why don't they crash into one another? If the Moon doesn't crash into the Earth, it's because the Moon circles the Earth: centrifugal force, which results from its movement, offsets the force of gravity. There is a balance between the two forces. The same applies to the Sun and the Earth: as the Earth goes around the Sun, its movement counterbalances the pull of the Sun. But to come back to

Newton, he never figured out why the stars don't crash on one another.

And why don't they?

In Newton's day, scientists were not aware of the existence of galaxies. Today, we know that the solar system circles the center of our Milky Way. It's this movement that keeps it in orbit and prevents it, as it prevents hundreds of billions of other stars, from falling into the central nucleus.

But what keeps the various galaxies from colliding with one another? As far as we know, there isn't any known center of the universe.

True. The response to your question lies in the expansion of the universe, in the general movement of the galaxies. We observe that these galaxies are moving away from one another. What the initial impulsion was we can only speculate about.

How long is this movement going to continue?

We don't yet know for sure. Imagine a rocket moving up in the sky above you. There are two possibilities: either it will fall back on Earth, or it will escape the pull of gravity from our planet. It all depends on the speed with which it was launched into space. If it was less than eleven kilometers a second, it will fall back; if it was greater, it will escape Earth's pull.

*

Would the same principle apply to the galaxies?

They are moving away from us, but their movement is slowed down by the gravity they exert on themselves. Their mutual pull on one another depends on their number and their mass, that is, on the density of cosmic matter: if this density is low, the galaxies are going to go on moving apart indefinitely (the scenario of the "open universe"); if the density is high, the galaxies are ultimately going to reverse their movement and begin moving back toward one another (the scenario of the "closed universe"). Those are the two possible futures of the universe.

And which of those two scenarios does science lean toward?

To date observations favor the first scenario (indefinite expansion). But this is not firmly established. We do know, however, that the expansion will go on for many hundreds of billions of years.

Scene 3

Earth

In the spatial desert, the first molecules begin an uninterrupted rondo and give rise to, in the suburbs of a modest galaxy, a very special planet.

The Crucible of the Stars

An infinite desert, with here and there clusters of galaxies fragmented into stars . . . A billion years after the Big Bang, the puree of matter has taken shape and is more recognizable. The whole picture appears stable, and the universe might very well have remained at this stage. But once again, evolution is going to intervene. Why?

Because the first stars are going to reverse the course of things. While everywhere else in the universe the cooling process is continually going on, the temperature of the stars is increasing considerably. The stars become the crucibles for fashioning matter and are going to be the cause of its taking a new step forward in cosmic evolution. The assemblages of the very first seconds of the universe are going to be replayed again in the stars.

In other words, the stars are in a certain way mini Big Bangs on a local level.

In a certain way, yes. The reheating is produced by the contraction of the star beneath its own weight. When the temperature reaches around ten million degrees, the nuclear force "reawakens." As was the case for the original Big Bang, the protons combine to form helium.

At its origin, the universe, one recalls, had stopped at this stage.

These nuclear reactions emit an enormous amount of energy into space in the form of light. The star shines. Our Sun has thus been "fueled" by hydrogen for four and a half billion years. The more massive stars shine much more brightly and use up their hydrogen in a few million years, at which point they begin to contract again. Their temperature rises to more than 100 million degrees. Helium, the product of hydrogen fusion, in turn becomes a fuel, at which point a set of nuclear reactions allows for new combinations: three helium nuclei are going to come together as carbon nuclei, and four helium nuclei as oxygen nuclei.

*

But why didn't these reactions take place at the time of the original Big Bang?

The encounter and fusion of three helium nuclei is an extremely rare phenomenon. It takes a long time for it to occur. In the original Big Bang, the phase of nuclear activity lasted for only a few minutes, which is too short a time to manufacture a meaningful quantity of carbon. This time, in the more massive stars, these agglomerations are going to take place over millions of years.

Each of these more massive stars is therefore going to manufacture carbon and oxygen nuclei?

For the next several million years, the centers of the larger stars will indeed be stockpiled with heavy nuclei, including carbon and oxygen. These elements are going to play a fundamental role in the following phase of our history. Carbon in particular, with its special atomic configuration, lends itself easily to the manufacture of long molecular chains, which will play a key role in the appearance of life. Oxygen will become a component of water, another element that is indispensable to life.

The Dust of Stars

And during this time, these stars continue to contract?

The heart of the star collapses into itself, whereas its atmosphere expands rapidly and becomes red, creating a red giant. When the temperature of such a star exceeds a billion degrees, it engenders heavier atomic nuclei, those of various metals—iron, zinc, copper, lead, gold—until it comes to uranium, made up of 92 protons and 146 neutrons, and even more. The roughly one hundred atomic elements that we know of in nature are thus the products of the stars.

That process could have gone on for a very long time.

No, because now the heart of the star collapses in on itself. The nuclei of the atoms at that point enter into contact and rebound. That provokes a giant shock wave, which results in the explosion of the star. That phenomenon is what we call a supernova, a burst of light that illuminates the sky like a billion Suns. The precious elements that the star has produced within itself throughout its long existence are at that point propelled into space at a speed of tens of thousands of kilometers a second—as if nature had taken the dishes from the oven at the proper moment, just before they would have burned.

But blowing up the oven at the same time!

*

That's the way massive stars die. Still, they leave on the premises, so to speak, a contracted stellar residue, which will become either a neutron star or a black hole. The small stars, such as our Sun, go out much more quietly. They exhaust their matter nonviolently and turn into white dwarfs. They cool off slowly and are transformed into nonradiant celestial corpses.

What happens to the atoms that escape from the dying stars?

They wander in interstellar space and mingle with the great clouds scattered throughout the galaxies. Now space becomes a chemistry laboratory. As a result of the electromagnetic force, electrons begin to orbit around atomic nuclei to form atoms. These atoms in turn combine to produce heavier and heavier molecules. Some of them contain more than ten atoms. The association of oxygen and hydrogen forms water. Nitrogen and hydrogen form ammonia. We have even found the molecule of ethyl alcohol, the component of our alcoholic beverages, which consists of two atoms of carbon, one atom of oxygen, and six atoms of hydrogen. All these are the same atoms that, later on Earth, will combine to form human beings. We are truly made of the dust of the stars.

The Cemetery of the Stars

At this point in the evolution of the universe, there is only gas, balls of stellar fire, but not yet any solid matter.

It's about to happen. As they cool down, certain atoms emanating from the stars, such as silicon, oxygen, and iron, are going to come together to form the first solid elements. These are tiny grains, smaller than a micron (one thousandth of a millimeter) in diameter, that contain hundreds of thousands of atoms. The force of gravity acts on the interstellar clouds and forces them to collapse into themselves, producing a generation of new stars. Some of these stars will have a suite of planets revolving around them, as is the case in our own solar system. And these planets will contain within them the atoms engendered by the dead stars.

In other words, for new stars to be born, old stars have to die. In space, too, the appearance of the new requires the death of the old.

The atoms of our biosphere have necessarily been created in the crucibles of stars and are sent forth into space when the stars die. These intertwined generations of stars and atoms begin to take place several hundred million years after the Big Bang and will go on for several billion years thereafter. Space becomes a kind of forest of stars: stars big and small, young and old, die, disintegrate, and enrich the terrain to nourish new growth. In our galaxy alone, an average of three stars come into being every year. Thus it is that, relatively late in the game, roughly four and a half billion years ago, one star of particular interest to us, our Sun, is born on the fringe of a spiral galaxy, the Milky Way.

Why "spiral"?

It's the rapid rotation of the stars around its center that gives our galaxy its shape, which is that of a flattened disk. The origin of the spiral arms is a result of complex gravitational phenomena. The Milky Way, that great luminous arc that crosses the sky at night, is the image of all the stars strewn the length of the disk of the galaxy and revolving around its center: our solar system makes a complete revolution around that center roughly once every two hundred million years.

An Ordinary Star

What distinguishes our Sun from the other stars?

In our galaxy, our Sun is an ordinary, run-of-the-mill star. Of the hundred billion stars, there are at least a billion that are so similar to our Sun you couldn't tell them apart. When our Sun came into existence four and a half billion years ago, it was much larger than it is today, and it was red. By slow degrees, it contracted, it became yellow, and its internal temperature increased. After a dozen million years or so, it began to transform its hydrogen into helium, like a giant H-bomb, with the difference that its output is controlled. This phenomenon of nuclear fusion is what assures our Sun's stability and its luminosity.

This run-of-the-mill star nonetheless managed to attract a number of planets and create a solar system around itself.

This phenomenon—stars with planetary systems around them—is probably rather common in our galaxy, although we've been able to detect relatively few because of the still-limited technology at our command. The formation of planets like Earth must be relatively recent. The solid bodies of our planetary parade are made up for the most part of oxygen, silicon, magnesium, and iron; the atoms were formed progressively by the activity of generations of successive stars. It took several billion years for them to come together in sufficient quantities in the interstellar clouds. We've been able to measure the age of the Moon, as well as that of many meteorites. The numbers are identical: 4.56 billion years, to be precise. The Sun and its planets appeared at the same time, at a point in time when our galaxy was already more than eight billion years old.

How did the planets form?

Interstellar dust gathers around embryos of stars and forms disks analogous to the rings of Saturn. Then, over long periods of time, these small objects come together to form rocky structures that become larger and larger. Some of the larger bodies attract other smaller bodies to them and eventually become planets. The countless craters of the Moon, and of other bodies in our solar system, attest to the violent shocks of the impact with interstellar matter, which added to their mass. These shocks emit a great deal of heat, to which is added the nuclear energy from the decay of radioactive nuclei.

So all these planets are in a state of molten incandescence?

At their birth, the large planets are incandescent balls of fire. The more massive the planet, the hotter it is, and the longer it takes to cool down. In the smaller bodies like the asteroids, the cooling process takes place very quickly. The planet Mercury and our Moon dissipated their initial heat into space over a few hundred million years. For a long time now, neither the Moon nor Mercury has had any internal fire and therefore no further geological activity. As for the Earth, this cooling process has taken longer. Today, the core of the Earth is still hot, and this causes the convective motion of still-fluid stone. These phenomena are what make continents shift, volcanoes erupt, earthquakes occur. This geological instability is actually a boon: it brings about variations of climate, which play a major role in the evolution of living things.

Liquid Water

What makes our planet different from the others?

Our planet has plenty of water in liquid form. There's water elsewhere in the solar system, plenty of it. The satellites of Jupiter and of Saturn contain water in the form of ice, because of their very low temperatures. Recent measurements by the *Galileo* satellite suggest that liquid water may be present on the surface of Europa. Venus also has water in the form of vapor, because, being second closest to the Sun, Venus has extremely high temperatures. The Earth's orbit keeps us just far enough from the Sun to allow water to remain liquid.

Mars also used to have liquid water, as the so-called canals and dried-up wadis that spacecrafts have shown seem to indicate.

As recently confirmed by the Mars mission *Pathfinder*, torrents of water did flow on the surface of Mars some billion years ago. But for a long time now, there has been none. Why? We really don't know. Given its relatively small mass, its tectonic activity is now very weak.

But where does Earth's water come from?

Let's go back to those torrents of matter projected into space at the death of the stars. Dust was formed—literally stardust—

on which ice and frozen carbon dioxide came to rest. When agglomerations of dust grew large enough to give birth to planets, the ice volatilized and escaped outside in the form of geysers. What's more, comets, which are made up largely of frozen water, fell on the planets, bringing water with them.

And the Earth retained that water?

Its field of gravity is strong enough to retain the water molecules on its surface, and its distance from the Sun allows it to retain water in liquid form, at least in part. In these early days of the Earth's formation, it was constantly bombarded by ultraviolet rays emitted by the young Sun.

The Gift of Water

Why didn't the same evolution take place on Venus?

We don't quite know. The two planets are so much alike they have virtually the same mass and contain the same amount of carbon. On Venus, however, this carbon is in the atmosphere, whereas on Earth, it is to a large extent in the ocean. Yet the atmospheric compositions of the two planets were very much alike in the early stages of their formation.

*

What does the difference stem from?

We think that water in the liquid state on the surface of our planet played a crucial role. Thanks to this blanket of water, the carbon dioxide in Earth's early atmosphere was dissolved and wound up at the bottom of the oceans in the form of carbonates. Venus is slightly closer to the Sun than we are. The difference in temperature was in all likelihood responsible for the absence of liquid water in that planet's early stages. Its atmospheric envelope of carbon dioxide created an enormous greenhouse effect, which kept its surface temperature in the vicinity of five hundred degrees. So it was that two planets, alike in many respects, evolved in two very different ways.

Without water—liquid water—it's safe to say that it would have been the end of our story.

I think so. Water played a primordial role in the appearance of cosmic complexity. Within the ocean blanket, sheltered from the ionizing rays from outer space, intense chemical reactions would occur. By means of various encounters and associations, those chemical reactions would produce molecular structures that were increasingly large. In the early stages of prebiotic evolution, carbon, born of the red giants, would play a major role.

An Atmospheric Face

Why is carbon so successful?

It's the ideal atom for molecular constructions. It has what we call a valence of four, meaning that it has four electron "holes" that can act as harnesses for numerous other atoms. The links it creates are sufficiently supple to allow easy and quick association or disassociation, which is indispensable to life. Silicon also has a valence of four, but the links it makes are much more rigid. It creates stable structures, such as sand, but it has no capability to yield to the constraints of metabolism.

It's therefore absurd to imagine that somewhere out there in the universe there is life based on silicon.

It's highly unlikely. In our galaxy, as in the neighboring galaxies, the various molecules of more than four atoms that we've been able to identify by radio telescope always contain carbon, never silicon. This observation strongly suggests that if life does exist elsewhere in the universe, it is also made out of carbon.

Once Earth's atmosphere was formed, life soon followed, isn't that so?

When Earth was born roughly four and a half billion years ago, the conditions were scarcely favorable. The temperature on the surface was too high. In addition, at that time, space was rife with countless small celestial bodies that would later be absorbed by more massive planets (the solar system was cleaning up its own house). The constant bombardment of meteorites and comets was extremely violent. Studies of various comets revealed the presence of a considerable quantity of hydrocarbons. The collisions of the first billion years in all likelihood brought, in addition to water, an important quantity of complex molecules to the surface of the Earth. These comets, which in ages past were generally thought to be harbingers of death and destruction, probably played a beneficial role in the appearance of life. Less than a billion years after the birth of Earth, its oceans were swarming with living organisms, including the first blue-green bacteria. This view was strongly confirmed by the rich harvest of organic molecules left behind in the tail of the comet Hale-Bopp in 1997, including formaldehyde, various cyanides, and methanol.

The Pregnancy of the Universe

End of Act 1, the longest and slowest. We arrive on Earth after several billion years of the history of the universe. From this point on, things are going to speed up considerably.

This time, the molecular agglomerations are going to take place with hundreds, thousands, millions of atoms. Ever since the

Big Bang, matter has been ascending the steps of the pyramid of complexity. Only a tiny fraction of the elements that have reached one step manages to ascend to the next. Only an infinitesimal portion of the protons in the first phase of our story has succeeded in forming heavy atoms. Only a very small number of simple molecules have organized into complex molecules, and only a tiny portion of these complex molecules will participate in the structures of life.

At the same time, it seems that there was a great degree of uniformity during the first part of evolution.

True, the universe has wrought the same structures everywhere throughout space. The fact is, we have never observed in the stars and in the most distant galaxies a single atom that does not exist in our own laboratories.

All of which suggests that the same story of earthly evolution could have occurred elsewhere, and that life could well exist on other planets.

We note that everywhere in the universe, quarks associate themselves into protons and neutrons, these protons and neutrons with added electrons come together to form atoms, and the atoms in turn form molecules. We also know that clouds of interstellar matter collapse to give birth to stars. We can well imagine that some of these stars do indeed have a suite of planets circling them, and we can also conceive that some of these planets may well contain liquid water, which is conducive to the appearance of life. All that is plausible. But as yet, we have no concrete proof.

The Earth In One Day

Time has also contracted: the further we go along in our story, the faster the evolution.

Absolutely. If we take the four and a half billion years of our planet and assume that it's but a single day—point zero of that day being the Earth's birth—then life begins about 5:00 A.M. and grows in complexity throughout the rest of the day. About 8:00 P.M., the first mollusks appear. Then the dinosaurs appear at 11:00 P.M. and disappear at 11:40 P.M., leaving the field open for the rapid evolution of mammals. Our ancestors put in their appearance at about five minutes to midnight; the capacity of the human brain doubled in the minute from 11:59 P.M. to midnight. The Industrial Revolution began in the last hundredth of a second.

And the world is filled with people who are firmly convinced that what they've been doing since this fraction of a second will last indefinitely. One can't help but see a logic in the unfolding of this first act, a kind of thrust toward complexity that propels the universe toward successive organizations, one inside the other, like so many Russian dolls, from chaos to intelligence. One might go so far as to say a sense, a meaning . . .

All evidence indicates that our universe has transformed its initial amorphous state into a variety of structures that are increasingly organized. This metamorphosis might be explained by the action of the forces of physics on matter that is cooling off. Without the expansion of the universe, without the great interstellar void, this story would have had no second act. But that only moves the interrogation back ever so slightly and brings us back to our reflections about the laws of nature. The question "Why are there laws rather than no laws?" strikes me as being a logical sequel to Leibniz's "Why is there something rather than nothing?"

Was the appearance of life programmed into the original scenario?

In the past, people used to say that the probability of life appearing was as unlikely as putting a monkey in front of a typewriter and expecting it to produce the complete works of Shakespeare. Today, there are a number of reasons to believe that the appearance of life on an appropriate planet is far from improbable. However, whether probable or improbable, we can say with certainty that from the very first moments of the cosmos, the possibility (but not the necessity) of the appearance of life was inscribed in the very form of the laws of physics—a fascinating question that Joel de Rosnay will explore in the next act of the cosmic drama.

Act 2

Life

Scene 1

The Primitive Soup

Neither too close to nor too far from an opportune star, the Earth takes refuge behind its atmospheric veil and, taking up where the stars left off, causes matter to evolve.

Life Is Born from Matter

DS: The idea that there is a continuity between the evolution of the universe and the evolution of life is recent. For centuries, we were very careful to separate matter from living things, as if they were two completely different worlds.

JDR: Life is capable of reproducing itself, of utilizing energy, of evolving, of dying. Matter is conversely inert, motionless, incapable of reproducing itself. When looking at the living world on the one hand and the mineral world on the other, it was difficult *not* to see them as opposites. But in the past, people had no idea that molecules were made of atoms, nor that cells were made of molecules. So the only way to explain the appearance of life on Earth was either through divine intervention or as a result of some extraordinary stroke of luck. It was, in fact, a way of hiding one's ignorance.

So in this second act, can I safely assume that we can rule out chance, an extraordinary stroke of luck, as you put it?

Not all that long ago, there were still some scientists who held on to the belief that life began through what they termed "creative chance." According to them, in the early period of Earth's evolution, certain chemical substances came together accidentally to give rise to the initial organisms—which implies that this "miracle" occurred only on Earth. Today, that hypothesis no longer has any validity.

Can we therefore assert that life was born of matter?

During the last few years, numerous discoveries and experiments have confirmed this major idea, which was first propounded in the 1950s: life resulted from a long evolution of matter that has been taking place on Earth ever since the first agglomerations of the Big Bang—from the first primitive molecules to the first cells, vegetation, and animal life. This forward thrust of living things, which has been going on for several hundred million years, is therefore clearly a stage of the same story, the story of increasing complexity. Following the birth of the Earth, molecules came together to form macromolecules, which went on to form cells, which in turn formed organisms. Life resulted from the interaction and interdependence of these new constituents.

Necessity, Not Chance

Would it be fair to say, as Hubert Reeves suggests, that the appearance of life was completely probable?

The French biochemist Jacques Monod went so far as to refer to the appearance of life on Earth as a "necessity"; given the conditions under which the Earth began, the laws that organize matter necessarily give rise to systems that are increasingly complex. If we compare the emergence of a living organism to a stone, it's easy to see why people found the appearance of life improbable. But if you take the long view, if you look at it over the full length of our history, it doesn't seem so far-fetched.

All of which suggests that the scene we're about to describe could well be taking place elsewhere in the universe.

Absolutely. Imagine for a moment a planet anywhere in the vast universe that is located at precisely the right distance from an appropriate star so that the conditions for life are possible. Imagine then that this planet is large enough to retain a dense atmosphere made up of hydrogen, methane, ammonia, water vapor, and carbon dioxide. And then let's picture that the cooling process taking place on this planet provokes an internal release of gas and a condensation that results in liquid water. Now make the reasonable assumption that the chemical syntheses that are taking place in this planet's atmosphere help that water accumulate molecules that are protected from ultraviolet rays. There's nothing exceptional about this whole set of conditions, which could be present in any number of places in the universe. And given these conditions, there is no earthly reason—no *celestial* reason—that systems of life should not evolve. That's why a number of scientists, including Hubert Reeves, believe that life could well exist elsewhere, either in our own galaxy or in some other.

Necessity, not chance.

Yes. Any planet that contains water and is situated at an optimal distance from a warm star has the possibility of accumulating complex molecules and small globules that will exchange chemical substances with their environment. From necessity to necessity, chemical evolution results in rudimentary living beings.

Recipe for Making a Mouse

Life that arises from inorganic matter is a little like what people used to refer to as "spontaneous generation." So our ancestors were not completely wrong.

That's true. But they thought that life occurred spontaneously, just like that, from matter that was decomposing. They theorized that earthworms came from mud, that flies were born from rotten meat. In the seventeenth century, one well-known physician even went so far as to give the recipe for making mice: take several grains of wheat and one soiled shirt well impregnated with human sweat, place both ingredients in a wooden box, and wait for twenty-one days. Simple, no? Then, when the first microscopes were invented, scientists discovered the existence of tiny organisms—yeast, bacteria—that proliferate in decaying substances, at which point it was categorically stated that life was constantly emerging from matter in microscopic form.

An idea that was not completely stupid.

The basic idea was correct, but the reasoning behind it was false: life does not evolve spontaneously; it took a very long time for it to appear. In 1862, Pasteur showed that microbial germs are present everywhere, not only in the air but also on our hands, on our bodies, on all objects. The microscopic organisms that we observe in the laboratory are therefore the result of a contamination. Pasteur concocted a broth of beets, other vegetables, and meat. He enclosed it in a flask with a very long, swanlike neck in order to isolate the mixture from the outside air, then boiled the mixture in order to sterilize it. No sign of new life ever appeared in his retort.

In other words, there seemed to be no way that life could appear spontaneously.

Right. But by conducting that experiment, Pasteur sent the problem of the origin of life back into limbo, where it remained for a long time. Because of him, scientists concluded that life could not come directly from inert matter; therefore, it could come only from life itself. Which raised the essential question: how do you explain the initial manifestation of life? There were only three solutions: divine intervention, which removed the matter from the hands of science; chance—in other words, some kind of accident—which took the matter into the realm of miracle, which is difficult to accept; or an extraterrestrial origin—germs of life that were brought here by meteorites—which didn't solve the question either.

Darwin's Intuition

And science nonetheless set out to establish a bridge between matter and life.

Yes. We had to get around the roadblock Pasteur had inadvertently set up and understand that life had evolved out of inert matter not spontaneously but gradually, over billions of years. It was Darwin who propounded this fundamental theory: the idea of continuity.

But he was talking about the evolution of the animal species.

Not exclusively. To be sure, Darwin discovered the principle of the evolution of living species: from the first cell to man, animals descend from one another by changing over the course of time by successive variations and by natural selection. But—and this we often forget or overlook—he also suggested that even before the appearance of life and the birth of the first cells, primitive Earth must have known an evolution of molecules.

What extraordinary intuition!

True. He even understood why it was difficult to prove that assertion and to observe it in nature: if molecules capable of evolving existed today in some little stagnant pond, they would come to grief, because today's living species would destroy them. That's a very prescient judgment: once life appeared, it in effect invaded everything, it consumed its own roots and kept other types of evolution from taking place simultaneously.

The Chicken and the Egg

How then do we prove that life "descended" from matter?

By retracing this evolution in the laboratory. We now know virtually all the stages that led from the molecules of primitive Earth to the earliest forms of life, and we can partially reproduce them in our test tubes. At the end of the nineteenth century, one researcher created a sensation when he succeeded in creating urea, one of the components of life, which is made up of carbon, hydrogen, oxygen, and nitrogen. But the discovery did not suffice to destroy the old prejudice, according to which life could be born only from life.

The story of the chicken and the egg.

Precisely. This vicious circle was finally broken by two researchers, Soviet biochemist Alexander Oparin and Englishman John Haldane. The conditions on primitive Earth, they maintained, were very different from what they are today: the atmosphere contained neither nitrogen nor oxygen but was made up of an inhospitable mixture of hydrogen, methane, ammonia, and water vapor—a grouping that was conducive to the appearance of complex molecules. During the 1950s, French paleontologist Pierre Teilhard de Chardin, another pioneer in this field, took up Darwin's idea about the evolution of matter and spoke about a "pre-life," an intermediate stage between inert matter and life itself, that could have taken place in the early period of Earth's evolution.

An idea that still remained to be proved.

Which happened very quickly, in 1952, when a young American graduate student, Stanley Miller, who was then only twenty-five, posed the question: why not reconstitute in the laboratory the same conditions that existed before life began? Since he was concerned that his colleagues would make fun of him, he conducted his experiments in secret. In his apparatus, he put the substances of the Earth's early atmosphere: methane, ammonia, hydrogen, and water vapor. Spark discharges introduced into the mixture provided energy and simulated the lightning of early Earth. Cooled water vapor dripped into the bottom trap, bringing with it newly formed compounds. The drops represented rainfall, the trapped water the ocean. After a week, analysis of this liquid showed that it contained simple organic molecules of low molecular weight, some of which are the components of the larger organic molecules of life (e.g., proteins, lipids, carbohydrates). What Oparin, Haldane, and others had proposed was given credence by Miller's experiment. The entire scientific community was in a state of shock: we had just built the first bridge between inert matter and life.

The Planet of Daisies

So it took some time before the scientific community was able to accept this notion of continuity between the universe and life. The task then became how to retrace the key stages in this evolution.

Three different sciences made the effort: chemistry, by simulating in the laboratory the principal transformations; astrophysics, by searching throughout the universe for traces of organic chemistry; and geology, by looking for the fossils of life on Earth. All that combined research led us to propose that the first components of life resulted from combinations of certain simple molecules that existed on Earth when it was formed, roughly four and a half billion years ago.

The chemical cocktail of primitive Earth, its liquid water, its special atmosphere, have all benefited from its proximity to the Sun. We were just the "right distance" from the star, they say, which doesn't tell us very much.

What that "right distance" means is that we're near enough to receive the Sun's infrared and ultraviolet rays to trigger chemical reactions, and far enough away so that the products manufactured do not burn up. In fact, that "right distance" is a way of talking about the equilibrium that was established on Earth during the period of its evolution. Let's imagine for a moment, to borrow a simple analogy from English physicist James Lovelock, a little planet populated by white daisies and black daisies. The white daisies reflect the light of the Sun and tend to cool the temperature of their environment. The black daisies absorb the solar light and warm their immediate area.

In other words, they're in competition.

Exactly. At first, the planet is extremely hot. The daisies can't cope with the high temperatures and die off in large numbers. But a few white daisies, grouped together in a specific area, cool their environment by their simple presence and manage to survive. The more the temperature decreases in their immediate vicinity, the more they proliferate and expand their growing area. After a certain period of time, they occupy almost the entire surface of the planet, which becomes essentially white. But then all of a sudden, the cooling process they've brought about drops the temperature to a point where the white daisies can no longer prosper, and they begin to die off in great numbers. At which point the black daisies, a few of which have managed to survive, have the advantage: by reheating the area in which they dwell, they get control. The system takes off in the other direction, until such time as the planet once again becomes too hot.

And that process can go on, I assume, forever.

No. With the passage of time, through an interplay of birth and death, an equilibrium is established: the planet becomes a kind of black-and-white quilt, wherein an optimal temperature is established at which both species can survive. The sets of surfaces—part white, part black—act like a thermostat. If for one reason or another the planet begins to overheat, the system brings the temperature back down until the opposite swing of the pendulum comes into play. After a certain period, the system becomes stable within relatively sustainable limits.

The Dawn of Life

How does that story relate to primitive Earth?

The story of the white and black daisies is the story of life on Earth. If the distance between the Sun and the Earth strikes us today as being just the "right distance," it's not because of some stroke of good luck but because the first components of life adjusted the temperature to the level most compatible with their survival and proliferation.

A kind of self-regulation. How did the various components actually do that?

Let's go back to the dawn of Earth, roughly four billion years ago. At this point, our planet possesses a core of silicates, a crust of carbon, and an atmosphere made up of the gaseous mixture of methane, ammonia, hydrogen, and water vapor. As a result of the ultraviolet rays emanating from the Sun and the powerful bolts of lightning striking the Earth's surface, these molecules of gas that are floating around the planet break up into pieces, dissociate, and regroup into more complex elements: the first molecules that we call "organic," because they are the components of what we today call living beings. For example, the atoms of carbon, nitrogen, hydrogen, and oxygen, which until then had been associated as methane, ammonia, and water, now come together to form amino acids.

Hubert Reeves has already noted the importance of carbon in the story of evolution.

Carbon in fact possesses a geometry that enables it to combine in a number of ways with other atoms to form stable structures, very reactive molecules, or long organic chains. It also has the ability to take electrons from one end of these chains to the other, which in a certain way can be thought of as the harbinger of nervous systems or of the electronic communications invented recently by man. The molecules of living things, therefore, are mainly combinations of atoms of carbon, hydrogen, oxygen, and nitrogen—and little more. As soon as these molecules gather in the atmosphere, they rain down into the waters of the ocean, where they're protected from ultraviolet destruction.

How long does this process go on?

For 500 million years. Thus, even in this very early period of the Earth's evolution, two essential characteristics of the living world were determined: its chemical composition—all organisms are made up basically of carbon, hydrogen, oxygen, and nitrogen—and its source of energy, the Sun.

The Organic Showers

Did other such showers take place on other planets?

As Hubert Reeves has already indicated, astrophysicists have discovered the existence of organic molecules virtually everywhere in the universe. Over the past fifteen years, they have identified about seventy-five such molecules, which shows that what happened here on Earth was in no way exceptional. Going back again four and a half billion years, there's every reason to think that these same organic molecules were forming throughout the universe.

We can therefore think of the initial elements of life, in a way, as raining down from the sky.

Yes. In the continuous shower of molecules that rained on the Earth there were amino acids and fatty acids, the precursors of proteins and lipids respectively. Two molecules, formaldehyde and hydrogen cyanide, seem to have played an important role at that time: subjected to ultraviolet rays, these two gases give birth to two of the four nitrogen "bases" of what will later make up

DNA, the basis of heredity. In this enormous culture medium of planet Earth in its early stages, two of the four "letters" of the genetic code that characterizes all living things already existed.

But in the initial chaos following the Big Bang, everything was all blended together.

It was indeed a kind of soup, as Hubert Reeves and others have described it, made up of very diverse molecules. And in that soup these new letters are now going to come together to form words, the amino acids, which in turn are going to combine by the hundreds to form sentences, the proteins. This time, it's the molecules that will carry on the work of complexity.

What could have happened to prevent these first syntheses from occurring?

Life itself, if it had existed earlier. Or heat and the ultraviolet rays, if they had been too intense. The reducing atmosphere of the Earth not only engendered these complex molecules but also protected them by acting as a cover, a shield. Later on, some early cells will make use of the Sun's energy to produce oxygen, and oxygen, in turn, will go on to form the ozone layer in the upper atmosphere, which will help protect the organisms from the ultraviolet rays. In other words, life was responsible for assuring its own survival.

Scene 2

Life Gets Organized

It's raining on the planet. Fallen from the sky, subtle molecules combine in the waters of the Earth and invent the first drops of what life is made of.

Born of Clay

Until now, our history looks very much like a Lego structure: the combinations are more and more complex and now form giant molecular chains. But we're still dealing with matter. What is the magic wave of the wand that will bring forth life?

A new stage of development can take place only to the extent that these molecules are capable of combining and recombining. In the universe, temperature was the triggering factor.

Here on Earth, it was a particular environment that would be responsible for the appearance of life.

You mean the oceans?

No. Life probably did not start in the oceans, as we have long believed, but more likely in tidal pools and stagnant ponds, places that were warm and dry during the day and cold and humid at night, places that alternately dried out and became moist again. In places such as these, there were deposits of quartz and clay, in which these long chains of molecules found themselves trapped. And there they began to combine with one another. Recent experiments, in which we've simulated this process of alternate dampness and drying out, have confirmed this theory of how life began. In the presence of clay, these well-known "bases" combine spontaneously into small chains of nucleic acids, which are simplified forms of DNA, the future prop of genetic information.

Life evolving out of clay! As was true for the origin of the universe, we find an amazing similarity between these pronouncements of science and those of early religions: in any number of ancient mythologies and religions, the origin of life is linked both to water and to clay.

Indeed, we scientists cannot help but note that coincidence. In those early myths, man—and woman—was fashioned by the gods from statues made of clay and water. Is that a simple coincidence, or an a posteriori statement? Human thought, like that of children, often possesses the power of simple intuitions, which science might later go on to prove.

The Invention of the "Within"

How did the clay react on the molecules?

Clay acts like a little magnet. The clay's ions, that is, its atoms or groups of atoms that either have lost their electrons or have too many, attract the matter around them and incite it to react. The well-known trace elements of today are in fact the result of the evolution of these little ions of the primitive ocean. Thanks to these ions, the combinations of matter can continue. Clay also acted as a desiccator, removing water so that larger molecules could be formed from their smaller building blocks.

To form once again the long chains of atoms?

Not only that. This time, a new phenomenon is taking place. Certain molecules are hydrophilic, they are attracted by water; others are hydrophobic, they are repelled by water. The proteins that exist in the tidal pools are made up of amino acids, some of which like water and some of which do not. What happens?

They form clusters, which puts their exteriors in contact with the water while the interiors are kept away from it.

In other words, they curl up?

In a way, yes. They close up, forming spheres. Other chains of molecules also form membranes and transform themselves into globules, which, at that point, appear in the oceans like drops of oil in a vinaigrette sauce. The appearance of these different pre-living globules is a fundamental problem.

Why is that?

For the first time in our history, something appears that is enclosed, that has an inside and an outside, as Teilhard de Chardin put it. This "inside" is going to preside over the next phase of the evolution of our little globules, right up to the birth of life and, later on, the birth of consciousness.

Consciousness evolving out of the magic of a vinaigrette!

In any event, life is born of the emulsion. What's interesting about these droplets is that they form closed environments, isolated from the primitive soup. They keep custody of the chemical substances that make up their own specific cocktails. They become, in other words, the new crucibles for the formation of life.

*

They take up the baton of the relay race of life, as the stars did at a certain moment in the first act, acting as spurs to bring about complexity.

Precisely. Without these membranes, new combinations would not have been able to emerge, a little like a human being without a skin. The establishment of closed environments was indispensable for this evolution to go on.

How do we know that?

We can easily duplicate that process in the laboratory. We take oil, sugars, and water and shake them up together, and we get emulsions made up of droplets that, seen under a microscope, look like cells. It's a completely spontaneous phenomenon. In the primitive soup, the molecules were large enough to come together, form an enclosure, and create these droplets.

And that phenomenon was taking place throughout the planet?

Yes, wherever there were tidal pools. The drops are all the same size, which corresponds to an equilibrium among their volume, their weight, and the resistance of their membrane (if their volume were too great, the membrane would break up).

The Drops of Life

But these droplets are not truly "alive," are they?

Not yet. Let's call them "pre-living." At this point in time, they are proliferating in extraordinary numbers. They have the advantage of being semipervious: they allow certain little molecules to pass through, which, once inside, are transformed into large molecules and find themselves trapped. At which point, a new alchemy gets under way, chemical reactions take place.

Each one of these drops concocts its own little interior? In a way, isn't that the start of individuality?

Indeed it is. And this phenomenon is going to bring about an enormous diversity in these pre-living systems. At times, the internal chemical cocktail causes the membrane to burst, and the molecules inside disperse. At other times, the chemical cocktail helps reinforce the membrane and in this way assures the survival of the system. Thus a kind of "drop selection" gets under way, which will last for millions of years. Before life begins, there is a struggle for life.

In other words, even at this very early stage, natural selection is already taking place.

Exactly as Darwin spelled it out. The drops that were best able to survive had an internal chemistry that their environment did not destroy. Those drops that had the best enzymes—the proteins, for example—had a distinct advantage.

What did these enzymes do?

Certain of them could derive energy from carbohydrates that these droplets "ate" from the ocean soup. All living things need energy for their cell work, and these pre-cells could use energy and continue on the road to complexity. Incidentally, these energy "producing" reactions were simple fermentations—not very efficient—but a giant step up for the cells-to-be. Later on, other drops that managed to retain pigments, that is, molecules capable of trapping light, transformed the Sun's photons into energy, like solar cells. Thus they did not have to "eat" to live.

Was that better?

Of course! Because the primitive soup, populated by all these bulimic droplets, began to become impoverished over time. The little structures that were autonomous had a leg up—so to speak—on those that needed to absorb substances that were becoming increasingly rare.

In that primitive soup, it's a bit difficult to think of anything being rare.

Yet such was the case. But all this would have led to nothing if another phenomenon had not occurred at this juncture: certain drops were capable of reproducing their little internal cocktails, which would give them a considerable advantage in the evolutionary process.

Survival Assured

How did the reproductive process come into play?

These particular drops contained a special chain of molecules, an acid called RNA—ribonucleic acid—which is made up of four molecules (almost exactly like those of future genes). Recently it has been proved that RNA possesses an extraordinary power: the ability to reproduce itself. Imagine that a drop breaks into two, and that the new drop possesses RNA identical to that of the first drop. Imagine too that this RNA plays a catalyzing role in the structure of the drop. There would thus be a transmission of a kind of primitive plan that could be useful for reconstructing a membrane and a system identical to that of the original drop. That, at a primitive stage, is an autoreproductive system. You can well imagine that these drops possessing RNA are assured of their "species' " survival.

Is it safe to say that these are the "first drops of life"?

Generally speaking, we define a living organism as a system that is capable of assuring its own preservation, of managing itself, and of reproducing itself. These are the three principles that characterize the cell, the basic structure of every living thing, from bacteria to man, and we can indeed attribute these principles to the primitive globules. If any one of these properties is missing, the organism is not "alive." A crystal, for example, is not alive; it reproduces itself, but it does not manufacture energy.

Is a virus alive?

A virus is more ambiguous. Take the tobacco mosaic virus, for instance, which attacks the tobacco plant. You dehydrate the virus to obtain crystals, which you can preserve in a wide-necked jar, the way you keep ordinary salt or sugar in the kitchen, for years. The virus remains still and does not reproduce itself, nor does it assimilate any outside substance. It's not "alive." It's a crystal. And then one day you take the powder and add some water. If you put a bit of the solution on the leaf of a tobacco plant, the plant will quickly show signs of infection. The virus has recovered its powers, using the leaf cell machinery, and it is reproduced with alarming speed.

So, is a virus alive or not?

Let's say that it's at the frontier of life. A virus is a kind of parasite that needs life to reproduce. It uses the cell like a copying machine. There was a time when we thought that viruses were the simplest form of life, and even that they were at the origin of life. But we no longer think that, the reason being that viruses need a living structure in order to be reproduced.

Today, we think that viruses are imperfect structures, the descendants of cells that in all likelihood evolved by ridding themselves of cumbersome material of reproduction in order to reduce themselves to their simplest expression and attain maximum efficiency. They apparently simplified themselves to attain the vital minimum.

Contamination by Life

Let's go back to our special droplets, those that can reproduce. From what you've said, I gather that they are going to start proliferating.

Within them, the game of chemistry is going on. The code of reproduction is improving. Perhaps by accident, a slight change in a base of RNA occurred. Two coils of this new nucleic acid came together to form a double helix, DNA, a structure that ends up asserting itself because it offers greater stability. At this point, a chemical dialogue gets under way between two types of molecular chains: proteins and DNA. In all likelihood, the reaction between the two was direct, one of the two fitting itself into the hollow of the other by a play of simple, regular chemical affinities.

So nature has reached the stage of genes, the underpinning of heredity?

*

The genes of all living beings on Earth are like segments of rosary beads twisted into a double helix and made up of four molecules, the four bases, like very long words written within the context of an alphabet consisting of only four letters. They fit together two by two, in a perfect equation.

So the drops with DNA are going to colonize the Earth?

In a way we can only describe as dazzling. The first droplets appeared on Earth roughly four billion years ago. In the following 500 million years, chemical selection was going on. It would seem that for a very long time, perhaps as long as several hundred million years, life remained dormant, limited to a few local zones of the planet, in tidal pools and ponds. And then, more recently, it suddenly invaded everything.

Over how long a period?

Perhaps over decades, maybe centuries. We're not sure. But it was a veritable explosion, especially if you compare it to the billions of preceding years. There was a geometric progression that began to take place, each cell dividing into two, then into four, then eight, sixteen, thirty-two, and so on. Very quickly, we reached astronomical quantities. At that time on Earth, there was nothing to stop their proliferation. Today, any new life-form would be wiped out immediately by the existing forms of life. Barely did it come into existence, so to speak, than life cut all the bridges behind it. In a sense, life contaminated Earth.

Is it possible to assert that there is a "logic" within nature that led it to find DNA and spread it throughout the Earth?

No. Nature doesn't "find"; nature selects. It proceeds by a process of elimination. DNA allows for a considerable variety of living structures. Those structures that, thanks to DNA, were able to reproduce naturally proliferated. And that's the reason DNA prevailed.

Would any life that might exist on other planets be founded on DNA?

In all probability. DNA is consistent with a logical chemical evolution of the universe.

The Red and the Green

How did our first cells evolve?

Within certain primitive cells, an early kind of respiration yielded the energy they required for their various cell functions. We call these "fermentation reactions"—the same as those that produce ethyl alcohol. This is not a very efficient way to get energy, but there was as yet no atmosphere of oxygen, and these early cells were not Olympians!

What is a more efficient way to get energy?

Two wonderful inventions are going to take place: photosynthesis, which produces oxygen gas as a by-product; and a truly efficient kind of respiration that uses this new environmental gas to release from carbohydrates a far greater amount of energy than does fermentation. Thus is born the early division of organisms into what will become the photosynthesizing plants of the world and those organisms using cellular respiration in the presence of oxygen: the future animal kingdom, as well as the fungi.

The Original Rift

The two worlds separated, but they remained dependent on each other.

Absolutely. In fact, they entered into a symbiotic relationship. By utilizing carbon dioxide and water, the photosynthetic cells produced oxygen (a waste product) and carbohydrates. If that sounds like "making a potato," it is just that—except that potatoes came much later with the advent of flowering plants. Our "potato" is a one-celled organism. The other "world" used oxygen in the breakdown of sugars (simple carbohydrates) to produce carbon dioxide and to synthesize energy storage chemicals, whose subsequent breakdown would yield energy to do the cell's work. A beautiful example of give-and-take.

These are the first "meals" of nature.

Exactly: the photosynthesizers were the makers of their "meals"; the nonphotosynthesizers were the "eaters"—eating to supply themselves with the requisite carbohydrates for cellular respiration and incidentally acquiring proteins, lipids, and other substances for cell structure and function.

These drops were now called cells?

Yes. And these primitive cells were going to continue their evolution by endowing themselves with a core, a nucleus. According to one theory, this new stage was the result of a strange coupling: plant life was thought to have developed from a host cell that had apparently taken on squatters—prokaryotes (plastids of a sort) that were transformed into chloroplasts (plant photosynthesizers). In a similar manner, the animal cell-to-be was apparently another host that cohabited with another type of squatter—bacteria with an especially good way of getting energy from sugars. These became mitochondria, a kind of microgenerator of energy that existed in all the evolved living cells.

*

A form of parasitism?

In a sense, yes. Or, more fairly, a symbiosis.

Were there any other possible evolutions for these cells?

Nature has doubtless known all the possible forms of reproduction and metabolism. It has sent out shoots in every sense of the term. But life such as we know it has eliminated all the other paths. On Earth, we know of another form of life that is extremely rare—namely, that found in the depths of the ocean floors, a life organized around the sulfurous resurgence of terrestrial magma. These are kinds of submarine oases where everything is red and orange. There, no green, no chlorophyll, can exist, because there is no light. Thus, sulfur bacteria start food chains: bacteria are eaten by micro cells, which in turn are eaten by micro fish, which in turn are eaten by larger fish, and so on.

The Colors of Life

In this story, nature never moves backward. It always forges ahead, always moves forward, increasingly toward the complex. Is it possible to say that nature possesses a memory?

There is a kind of chemical memory, in the sense that a molecule is simultaneously a shape and a piece of information for other molecules. These shapes are complementary, they fit into one another, they have affinities, they recognize one another. The molecular world is a world of signs, and chemistry is its language. Certain populations of molecules conduct energy, others are meant to reproduce, others are isolated from water, still others attract clouds of electrons. That is what pigments do, for example. Do you know why the world is colored?

Not only for beauty's sake, I would imagine.

Not for that alone. A pigment is a molecule that possesses extremely mobile electrons. This characteristic enables it to absorb packets of light—the photons—to reflect certain spectra and therefore to color matter. But at the same time, it encourages the construction of molecular chains that go into the building of living things. Pigments preside over a subtle chemistry that does not require a great deal of energy. It is because both hemoglobin and chlorophyll are endowed with these properties that they are so important to the composition of life.

Beauty above all. In other words, there is no way that the living world could ever have been gray.

Probably not. Nor could it have been all white or all black. Color is intimately associated with life.

False Coincidences

Once again, time played a fundamental role in this phase of our history.

Indeed it did. Time contracts or grows longer according to the phase of evolution. The appearance of a very reactive molecule concentrates space-time: it can invade its environment and neutralize, in the space of a few seconds, other molecules that have taken several thousand years to evolve.

From primitive Earth to the first cell: can we say that the scenario is now complete?

We know the main stages, but a few gaps in our knowledge remain. What we don't yet know for sure is just how the reproductive mechanisms came into play. Some researchers still hold on to the idea that life began outside our planet and that it was brought here by a meteorite that "contaminated" the planet. Actually, it's not such an outlandish idea.

Can we reproduce this evolution in the laboratory, by chemical syntheses, and produce life in test tubes?

Almost. A number of scientists are working on that right now. It's a new area of science that is dealing with what we call "artificial life," which consists of several approaches. We can produce syntheses of molecules, or we can give rise to spontaneous evolution in our test tubes, by creating conditions of Darwinian selection to produce molecules that are capable of reproducing themselves. We can also leapfrog over several stages by utilizing computer simulation. Today, science has even managed to create robot-insects that are capable of adapting spontaneously to new situations: these "creatures" can climb stairs, get back up when they fall, flee from heat, and emit signals among themselves. Some researchers are even trying to create other forms of life, for example, using silicon as their basis.

One can't help but note in the context of this biological history that, as was the case for the evolution of the universe, there seems to be an innate logic to it. Would you go so far as to say, as Nobel Prize laureate François Jacob once suggested, that life evolved in a logical sequence?

Let's just say that a succession of chemical reactions led to irreversible situations and new properties. All this constructed a story, and we humans found ourselves at the end of it. We then set about retracing our evolutionary steps. We find our evolution unique because it is our own.

Still, you have to admit that there are an amazing number of coincidences in that story of life.

They aren't really coincidences. Look at it this way: let's say a soldier is telling you about some extraordinary event that happened to him during a war. He was in an apartment; a bomb hit the roof and exploded. By some miracle—or coincidence—he took refuge under a bed, which saved his life. Or another: he parachuted from a plane, and his chute failed to open all the way. He was plummeting toward the Earth, twisting like a corkscrew as he fell, and by some stroke of luck he landed in a swamp, which cushioned his fall and saved his life. If his stories sound incredible, it's simply because he survived to tell the tale. There were millions of other soldiers whose stories ended tragically, but unfortunately, they aren't around to tell us what happened. Life is like that. If it seems to be the result of an incredible number of coincidences, it's because we're ignoring the millions of different paths that led nowhere. Our story is the only one we're able to reconstruct. And that is why it strikes us as so extraordinary.

Scene 3

The Explosion of the Species

Cells, too long solitary, become interdependent. A world rich in color begins to flower. Species come into being and diversify. Life grows and multiplies.

The Solidarity of Cells

At this stage of our story, Earth is inhabited by cells that dwell peacefully in the oceans and could just as well have gone on living as they always had.

But there came a time when these cells were obliged to evolve. The first cells, which were proliferating, began to poison themselves with the waste they were spewing back into the environment. From the very beginning, life showed a natural tendency to form groups of individuals. These cellular "societies" have obvious evolutionary advantages. They are better protected, and they have increased chances for survival than do isolated cells.

How do they go about getting together?

The behavior of an amoeba, the myxamoeba, which still exists today, can help show us how. This amoeba feeds on bacteria. If you deprive it of food and water, it emits a distress hormone. Other amoebas rush to the rescue and gather into a colony about a thousand strong—a thousand "individuals," as it were, moving like a slug in search of nourishment. If they don't find it, they stop moving, put up a spore-producing stalk, and remain there indefinitely, just like that, as long as it's dry. But if you add water, the spores germinate and give rise to independent myxamoebas, which head off in different directions. Volvox provides another example. A colony of biflagellated cells forms a hollow sphere. A single cell cannot survive outside the colony; its cells are beginning to be interdependent and specialized. For example, only certain members of the group can reproduce. Also, there is intracellular communication that coordinates the flagella so that movement in a specific direction can occur. This is a giant step on the road to multicellularity.

Is this what you would describe as the first multicellular organism?

It's highly likely that logical "agglutinations" actually occurred in the early phases of life. Some of the first associations of cells benefited from having a central tube, a kind of drainpipe that evacuated wastes. Other multicellular organisms were tapered and were endowed in the front with a system of coordination and at the rear or on the sides with a kind of propulsion system. They thus remained joined together.

What did these "packages" of cells look like?

They were made up of several thousand individual cells and formed the earliest marine organisms: first sponges, then little primitive jellyfish, and finally a succession of ever more complicated worms. This transformation took place over a relatively short span of time: a mere few hundred thousand years. Evolution is accelerating.

The Division of Work

These new agglomerations, I take it, were very different from their predecessors.

Absolutely. Matter is made up of groups of atoms that for the most part are identical to one another. In the living world, cells that come together are differentiated according to where they are located in the structure. Some of them would be specialized in locomotion, others in digestion, and still others in conducting energy. Little by little, as they reproduced over many generations, these organisms transmitted these properties to their offspring.

Is it safe to assume that this phenomenon, once again, can be ascribed solely to the urgent need to survive?

Yes. An organism made up of specialized cells resists better than an entity composed entirely of identical cells, for the simple reason that it can respond to the various aggressions of the environment in different ways, which gives it a greater chance to survive. Monolithic systems always disappear in the long run.

But what impels these cells to join together? Certainly not because they say to themselves: "We'd better join forces if we want to survive!"

Of course not. Cells obviously don't "know" that what they're doing is in their own self-interest. But they do possess a linking mechanism that, in a sense, "invites" them to hook up with their fellow cells, and when they do, there is an interplay that occurs: the cells exchange substances with one another. The give-and-take of this chemical communication, and the slight changes that affect their genes, ends up making them specialized. A topography is thus set up within the group of cells. A jellyfish, for example, possesses a system of contraction that enables it to move about and a sensory system that gives it the ability to direct itself toward sources of food. The blueprint of these abilities is contained within each individual cell. At some stage, there developed a single reproductive cell containing all the DNA information needed to form a new organism.

Despite all you say, cells that remained alone did survive, some of which still exist to this day. Why didn't they join forces like the others did?

Because they were, in their solitary state, already well adapted to their environment. Such was the case with the amoebas and the paramecia. They were protected by a fluid membrane, and the latter were also equipped with cilia capable of beating like oars in tandem rowing, which allowed them to move about at will. Others, like the euglenoids, were endowed with photosensitive spots that indicated where the light was, so they could photosynthesize. They also ate and had efficient enzymes that digested all sorts of prey. Euglenas are blessed with a kind of flair: chemical receptors communicate with their flagella and guide them toward richer sources of food, much the way our human sense of smell tells us where the meal is being prepared.

Long Live Sex!

How are multicellular organisms going to continue evolving?

Starting with the simplest multicellular organisms, such as algae, sponges, medusas, and the like, the tree of life developed into three main branches: the branch that contains mushrooms, ferns, mosses, and flowering plants; the branch of worms, mollusks, crustaceans, arachnids, and insects; and the branch of fish, reptiles, and prechordates, then birds, amphibians, and mammals.

And somewhere, sooner rather than later, there was a major "invention": sex. Until then, cells reproduced by dividing into two identical units. With the advent of sex, two living creatures came together and produced a third being that was different from the other two. What sly little fellow dreamed that one up?

According to some theories, sexuality was the result of cannibalism. By literally eating one another, cells integrated the genes of other species, which later on in the evolutionary process blended together. This phenomenon already existed among bacteria; some of them, baptized the "more" and the "less," paired off and exchanged their genetic material. Later on, as organisms became increasingly complex, they were endowed with cells that were specialized for reproduction, what we call germ cells, each of which comprises half the genes of each parent organism. Sexuality became increasingly common.

And from that point on, the living world became more and more varied.

It was a revolution. Thanks to sexuality, nature was able to stir up the gene pool. There was a veritable explosion of genes. The great adventure of biological evolution was about to begin. It was an adventure filled with countless failed attempts, paths that led nowhere, species that failed to survive. Nature was testing in every direction: if the newly invented species couldn't adapt, it disappeared.

Why was it determined that sexuality consisted of two sources? Why not three?

The molecular evolution that produced DNA shows us why—and how. By a special process of cell division that produces an egg or a sperm, each parent cell's DNA is precisely halved. In the fertilized egg, the two halves of DNA are joined together, providing the new organism with one half of each parent's genetic information. DNA would have to have a very different structure in order to provide only one-third of itself to a sex cell. Nature did not intend a ménage à trois.

The Necessity of Death

Another decisive phenomenon comes into play: the introduction of time in relation to the organism, that is, the fact of growing old and, at the end of a certain period, the disappearance of the individual. In other words, death. Wasn't there some way that death could have been avoided in the evolutionary process?

Death is as important as sexuality. "Ashes to ashes and dust to dust" is more than a religious incantation. When we die, we put back into the world atoms, molecules, mineral salts that nature needs to continue its evolutionary process. Nature is constantly engaged in a gigantic recycling process, whereby the number of atoms in play remains constant, as it has been ever since the Big Bang. Thanks to that process, animal life can constantly regenerate itself. The molecules of each of our bodies have surely been recycled countless times.

Was this aging process already present in the earliest organisms?

Yes. In all living beings, most cells constantly reproduce, but they possess a chemical oscillator, a kind of internal biological clock that limits the number of times they can reproduce, which is between forty and fifty. When they have reached that stage, a programmed mechanism within the genes leads them, as it were, to a kind of suicide. The cells die. The only ones that elude this fate are cancer cells: they reproduce indefinitely, without specializing or differentiating themselves from embryonic cells.

But these cells' "immortality" causes the death of the organism of which they're a part. Is it fair to say, therefore, that death is a necessity of life?

*

Absolutely. It's a logical consequence of living. To the extent that cells divide, they multiply the errors of their genetic messages that accumulate over time. Finally, there are so many errors that the organism deteriorates and dies. This is an ineluctable phenomenon. Death is, to be sure, not a "gift" for the individual involved, but it's the premium a species pays for having lived.

Once evolution has devised sex and death, what more, if anything, remains to be done?

Make organisms better, improve organisms. From differentiation of cells comes specialization of function, which leads to the development of tissues and organs, culminating in the most important improvement: complex organ systems at whose pinnacle sits the nervous system, coordinating all. Added to this is an immune system that protects against "foreign invaders" (parasites, bacteria, viruses, pollen, and so forth), plus a hormonal system of checks and balances controlling biological rhythms, including sexual reproduction.

When did the nervous system appear on the scene?

It began with multicellularity. Cells became increasingly specialized, and the need arose to keep their operations organized. First came cells (neurons) capable of transmitting a simple message—initially in both directions, but then in only one direction. The first neurons were poised for the great ascent to become the human nervous system, with its large brain. Along the way, neurons produced a nerve net; later, clumps of nerve cells formed ganglia capable of coordinating an area within the organism. Forward-moving animals demanded that sensory organs congregate in the head end, hence more ganglia formed there until they joined to make a primitive brain. And this brain was the commander of a vast system of nerves that had formed from single neurons.

The Gift of Tears

What prompted organisms to leave the water?

First in this migration were the plants. Since they could literally make themselves, it became a matter of adapting to terrestrial environments. First algae on the shores devised structures that allowed them to survive in a less watery home. Then followed mosses, ferns, and vascular plants, a veritable feast that beckoned to the increasingly hungry marine life. Certain marine invertebrates were the first animal invaders; they, in turn, added a new dimension to this terrestrial banquet. Primitive fish sought insects and, over time, developed stronger fins with which to move in their search, as well as an air-breathing lung, enabling them to remain out of water for longer and longer periods. With the passing generations, the descendants of that species managed to remain longer and longer on dry land, thanks to their well-developed lung, which could pick up oxygen from the air, and to their tears. It was absolutely essential that their eyes be kept moist with tears if they were to see as clearly in the air as they had underwater. Through successive selections, the species improved: their fins became more solid; they grew a tail. Their descendants became frogs and other amphibians. We wouldn't be here if that fish had not had tears!

Was life in the open air favorable to evolution?

Yes. In the open air, communication was more immediate, quicker, simpler. There was greater accessibility to food. And yet oxygen is a poison for life. It favors the birth of free radicals, unstable molecules that induce cellular destruction and therefore precocious aging. But oxygen was also essential in providing organisms with the energy they needed to move evolution forward.

How did these new constraints of the terrestrial environment speed up the perfection of organisms?

These newly evolving quadrupeds developed sturdier bony skeletons, to which was attached an improved musculature. In short, all their organs, with the exception of the reproductive system, adapted, resulting in better and better land animals. To their detriment, however, they had to return to the water to lay eggs, like their fish ancestors. The metamorphosis of the fish like tadpole into an air-breathing four-legged frog is a wonderful recapitulation of this saga. Reptiles overcame the handicap completely. They "invented" the shell so that their eggs could be laid on land, freeing them from the constraints of an aquatic environment. Of course, a reptilian explosion followed, as they migrated widely and changed to occupy the niches that their terrestrial environments offered. Pectoral and pelvic girdles improved, taking weight off their bellies. They ran, they climbed, they flew; some evolved into birds, others into the first mammals. All the while, the ever-crucial brain was enlarging to coordinate all their activities. Eventually, egg laying would yield to the mammalian internal fertilization and gestation. When huge numbers of eggs did not have to be laid to ensure a species' survival, vast amounts of energy could be diverted to a host of mammalian "improvements."

The Gift of Plants

Why didn't the fungi and the plant world develop all those wonderful things that were being invented in the animal world?

With the exception of algae, which evolved near the surface of the water, the plant world followed a much more economical path, thanks to its immobility, which allowed it to avoid spending too much energy. Its way of life was very simple: solar cells to transform the energy of the Sun directly into chemical energy, and roots to draw water and mineral salts from the Earth. The neatest trick with plants is their reproductive systems, which, because of their immobility, employ a variety of methods. The plant world also inherited a sexuality that is extremely rich and adapted itself beautifully. You only have to look at a mushroom growing at the base of a giant redwood several thousand years old to realize this amazing variety, or even the ordinary evergreens growing on the side of a mountain.

In what way are evergreens the result of good adaptation?

In the forest, trees need a certain temperature to develop. As was the case with the daisies of our imaginary planet, the dark-hued leaves are quick to capture even the weakest rays of the Sun, warm up their immediate environment, and create a microclimate favorable to their growth. But in winter, these same trees are covered with snow and become white. If they were to remain that way for a considerable length of time they would no longer be able to provide for these favorable conditions. But since their branches point down toward the ground and their needles are pointed, the snow doesn't stay on them very long and slides off, thus restoring the evergreens to their former color and enabling the trees to perform their warming function more quickly. Evolution has retained this kind of tree, which resisted widely divergent climatic changes. And that is why you find so many evergreens in the mountains and in colder regions.

*

And why we go into such raptures about the evergreens' wonderful adaptability. A naive question: why didn't the plant world develop brains, the way the animal world did?

Motionless organisms don't require functions of complex coordination. They are not prompted by the necessity to flee, to defend themselves, to fight, the way animals are. But don't assume the total passivity of plants. We're just beginning to learn a great deal more about them. For instance, we're just discovering that plants possess a kind of immune system, a means of communication, and even a system analogous to the animal world's nervous system. Plants possess sophisticated mechanisms that protect them against invaders: a kind of vegetal "hormone" that enables them to marshal their defenses. We also know that trees have a way of warning other trees about the presence of an aggressor.

"Warning"?

Absolutely. When they're in the presence of animal predators that want to eat their lower branches, certain trees emit volatile products that, transported from tree to tree, modify the production of proteins and give the leaves an unpleasant taste. I wouldn't go so far as to say that you ought to talk to the plants in your apartment, but . . .

Still, isn't it fair to say that the members of the animal world have progressed much further than those of the plant world as far as complexity is concerned?

*

It's true that in adapting to the environment, the animal world has gone further and faster than the plant world. There are species that run, burrow, dig, fly, climb. Animals have developed countless defensive tricks, from the pressure buttons of the June bug to the tentacles of the octopus. Animals invent lures, ruses, and weapons: claws, wings, beaks, fins, flippers, shells, tentacles, poison.

Natural Exclusion

You just used the term "invent." Is that an accurate description?

No, they don't "invent." I'm really referring to the phenomenon of "selection," which eliminates the least apt. Take, for example, the large-beaked sparrow, which feeds exclusively on tiny worms that nest in the cracks and crevices of trees. There were so many sparrows with normal, smaller beaks that before long they had eaten all the worms to be found on the bark of the trees. With nothing to eat, the majority of the species died off. But a tiny minority, thanks to a chance mutation, were endowed with longer beaks. These sparrows and their offspring were able to dig deeper into the cracks of the trees and find enough food to survive. The result: that branch of sparrow survived and multiplied so that over the course of many generations, most of the sparrows had longer beaks. It's not fair to say that the sparrow "invented" that neat little trick. In reality, it was the reverse: those that didn't have the good luck to have the mutation that gave some sparrows longer beaks, simply died off.

So there is no "intention" in the process of evolution.

No. In populations, there are many variations, some of which are advantageous and some of which are not. Those that allow for survival are, by definition, preserved in succeeding generations.

Doesn't environment act directly on evolution?

Today, we think that the environment may have an influence on cells, via the intermediary of the mitochondria—those little subfactories within the cells that possess independent genetic blueprints and are extremely sensitive to change. But we also know that that influence is not transmitted from generation to generation.

Is the principle of natural selection still relevant today?

Yes, provided you don't try to read into it the notion of an environmental god, a kind of creative deity that decides what is good and valid and what is not—this we keep, and that we throw aside. No. It's better to think of it in terms of competitive exclusion: with the passing generations, the less adaptive species are excluded. To really understand this process, you have to think of it over the very long term, as something taking place over many successive generations. The changes and modifications we are talking about occur very, very slowly.

Of all the species "invented" by nature, the overwhelming majority disappear. Aren't there times when evolution is tempted to stop, when the living world can find its stability, like the daisies on our imaginary planet?

No, because from the very earliest phases of life, the diversity is enormous. Harking back to Hubert Reeves's metaphor, there are simply too many evolutionary letters to form only one word. Maybe a few primitive species may have established a kind of stability on some tiny asteroid in a kind of compromise or armistice with evolution, but not here on Earth with its size, its geology, its relation between the mineral and organic worlds, and its environment in constant change, which compels species to modify their adaptations and to evolve.

And all that takes several million years?

Yes. This natural selection acts upon millions of successive generations. The sensory mechanisms become more and more refined, behavior more and more diversified. Certain organisms join forces and form a veritable collective organism. Take a beehive, for example: it keeps the hive at a certain controlled temperature by the collective movement of the bees' wings; the hive is irrigated by the hormones resulting from the insects rubbing up against one another. When scout bees leave the hive in search of food, they return to indicate the best sources by doing a kind of dance. In this way, the hive economizes its energy and, by doing so, optimizes its chances for survival. The same is true for the ant world: they take care of the larvae, they care for the queen, they have a distinct division of labor—in much the same way that the cells of the volvox do—and thus they maintain the equilibrium of the anthill. If you take away 30 percent of the ants in the anthill, the remaining 70 percent adapts and divides up the labor so that the proportion is reestablished.

But ants are incapable of acting alone, of having anything like autonomous behavior. Their behaviors are programmed in their DNA, in their genes.

And they are incapable of planning. Ants communicate individually by pheromones, but also collectively via the environment. A newborn ant quickly learns all the networks, the paths mapped out by his predecessors. The simultaneous behavior of thousands of individuals leads to a kind of collective intelligence. The anthill knows, for example, what the shortest path is between the anthill and the food source. This method of collective association can be rated as highly successful, for it has existed for millions of years. If there was ever a nuclear war, it is probable that the ants would survive, because of their chitinous outer skeleton, which enables them to resist radiation, and because of their highly developed methods of organization.

The Dinosaurs' Misfortune

A world of ants and bacteria . . . not exactly an exhilarating prospect! The more we go on in our story, the clearer it becomes that the evolution of the universe, like that of life, has been, to say the least, chaotic.

True. There has been constant acceleration, but there have also been crises, dead ends, and periods of great extinction. Two hundred million years ago, dinosaurs ruled the Earth. Never before had any species succeeded in conquering all the milieus they had. There were all kinds, sizes, and shapes: little ones, huge ones, vegetarians, meat eaters. Their relatives ran, flew, swam. An enormous diversity enabled them to adapt to their environments.

Yet they were going to vanish from the Earth? Is the hypothesis that their disappearance was a direct result of their inability to adapt totally wrong?

Totally. At the end of the Jurassic period, roughly sixty-five million years ago, an enormous meteor three miles in diameter crashed into the Gulf of Mexico not far from the Yucatán peninsula. The shock was such that it reverberated from one end of the planet to the other and caused a resurgence of magma, which, as you recall, is the molten rock deep within the Earth that gives rise to igneous—or burning—rock. As a result, there were forest fires everywhere; the whole world was on fire, and the burning forests released not only great quantities of carbon dioxide into the atmosphere but also vast clouds of ash, which covered the Earth in an immense veil, obscuring the Sun and causing the Earth's temperature to drop precipitously. After a period of terrible cold, there was probably a greenhouse effect, which led to the Earth's rewarming.

But only a few species survived this catastrophe?

Very few. A group of very small prosimian primates, the tree shrews, began developing in ways that were crucial to the rise of anthropoids and eventually ourselves. They were tree climbers, and grasping ability became advantageous. An opposable thumb, nails rather than claws, and sensitive pads at the ends of the digits resulted. Additionally, the long snout of their predecessors gave way to a more flattened face and forward-facing eyes as sight, now stereoscopic, rather than smell became the dominant sense. There was an ongoing increase in the visual centers and the cerebral hemisphere of the brain as these new skills were acquired.

Selection in the Head

At what point did the brain—the real brain—make its appearance?

Actually, the fused ganglia of invertebrates can be considered a brain. But if you mean the vertebrate brain, we can go back to jawless fish with skeletons of cartilage. Remember, as activities and skills improve, new structures evolve that require and result in bigger brains. A jaw, a bony skeleton, four limbs, improved muscles, refined senses—all these are associated with increased brain size and function.

Essentially all brains, from those of fishes to that of man, consist of a hindbrain, a midbrain, and a forebrain. The medulla, and especially its associated cerebellum, increases vastly as we proceed up this vertebrate ladder. The midbrain's changes are more of function than of size. For example, birds with their superior eyesight have larger optic centers than do mammals. The most dramatic increase in size and complexity occurs in the cerebrum of mammals, the seat of intelligence and of language.

The most amazing thing, it seems to me, is this principle of selection that you find everywhere in the universe: in the initial chemistry of the molecules, among all living creatures, and, if one is to believe a number of eminent neurobiologists, within the brain of a newborn child at the moment of birth.

The development of the nervous system would also obey the principle of Darwinian selection. When a young animal grows up, its neurons transmit by obeying a blueprint provided by the genes. But the linking of two neurons can occur only if they are both functioning in a circuit and if they are asked to do so by the environment. The visual neurons of a newborn child, for instance, would not connect if that child was kept permanently in darkness. So, in a certain sense, there is a selection that retains only the circuits that pertain to the individual. You learn by a process of elimination.

American anthropologist Stephen J. Gould believes that every event, however insignificant, influences the course of history. As in Frank Capra's classic film *It's a Wonderful Life,* all it takes is one minor change, an alteration in the slightest detail, and everything else changes with it, with a flood of consequences. If a certain worm from which we directly descended, had not appeared, or if the climatic changes had not killed off all the dinosaurs, we wouldn't be here today. So according to Gould, there is no sense to evolution. Evolution doesn't favor the species that adapts the best but the species that has the best luck. Life may have been a probability in the chain of evolution, but as far as mankind is concerned, isn't it fair to say that we're simply damn lucky?

*

If the climbing tree shrews had not survived at the time the dinosaurs disappeared, we wouldn't be here. There is no intention hidden within our story. But the end result is nonetheless that things do become more and more complex as we go on. If there are planets out there somewhere that have developed under the same conditions as Earth has, it is not improbable that there are living creatures on them, creatures that are no different from us than, say, an ostrich is from a crocodile: creatures with four limbs, two eyes, a brain, systems of locomotion. And there is a very good chance that these creatures are at roughly the same stage of evolution as we are. Although we cannot affirm that there is any law that moves us ineluctably in the direction of increasing complexity, we can assert that something is happening that is leading toward an intelligence that is both increasingly great and increasingly abstract. But perhaps the history of evolution is the artifact of a consciousness that is becoming aware of itself.

The Memory of Origins

Only the human brain asks questions about itself. Is that what distinguishes it from all the other species?

Not only that. Humans are able to exteriorize functions in the environment. The tool prolongs the hand. Man can now do what all the other animals do: with an automobile, he can "run" as fast as a gazelle; with a plane, he can "fly" like a bird; he can swim underwater like a dolphin, burrow underground like a mole. . . . A mask, glasses, a parachute, wings, wheels . . . Mankind has also extended the sensory functions through writing, which enables us to preserve the written word in both time and space. That is what characterizes the human brain: it's not simply a soft mass of neurons, not a telephone switchboard that brings together all the body's different circuits, nor is it a computer. It's a fluid network, constantly being reorganized, which reconfigures its neurons through both action and reflection.

Throughout this entire story, we note that complexity develops by the arrangement of simple things: quarks in the beginning of the universe, the symmetry and valence of carbon, four nitrogen bases for the genes, two very similar molecules found in both the plant and vertebrate worlds, two different sexes. It seems as if nature, at each stage, always chooses the simplest path in order to progress.

In a sense, yes. But complexity does not mean complication. It's a repetition of simple elements that reproduce and proliferate. Today, on a computer screen, we know how to simulate this process: starting with an elementary form or shape, we can construct elaborate drawings that are called by the lovely name "fractal forms." They look like butterfly wings, the tails of seahorses, like mountains and clouds. Life is repetitive. The atom is in the molecule, which is in the cell, which is in the organism, which is in society.

We therefore bear within ourselves the traces of these boxes within boxes.

Exactly. Our large, complex brain preserves the memory of evolution. So do our genes. And the chemical composition of our cells is a little piece of the primitive ocean. We have preserved within ourselves the milieu out of which we came. Our bodies tell the story of our origins.

Act 3

Mankind

Scene 1

The African Cradle

Clever little monkeys are born into a world of flowers. To combat the dryness, their descendants stand up and discover a whole new world.

A Barely Presentable Ancestor

DS: "If it's true than man descended from the monkey, at least let us hope that word of it doesn't get around!" exclaimed one proper English lady in 1860, upon hearing about Charles Darwin's theory of evolution. Unhappily for her, word has indeed got around.

YC: Not completely. You'd be surprised, but it's still difficult for many people to accept that fact, to admit to that relationship. The animal origin of the human race runs smack into so many religious and philosophical beliefs that many people simply prefer to deny it. My maternal grandmother, a lady of staunch principles, said to me one day, in all seriousness: "*You* may have descended from a monkey, but *I* certainly didn't!" There are many people who are still in a state of confusion on this point. When we say that we descended from a monkey, some people believe that we're talking about a chimpanzee.

Man didn't descend from today's monkey, but from a particular monkey, isn't that so?

Absolutely. We descended from a species that was the common ancestor of both lines: the line of superior monkeys of Africa, on the one hand, and that of the prehumans, then humans, on the other. Man therefore is not a monkey in the broad sense of the term, except from the viewpoint of his place in the general classification of animals. What makes us different has been our ability to transcend that simple condition. Joel de Rosnay has reminded us in no uncertain terms: there is no way we can ignore our affiliation. We bear it within our bodies.

It would even appear, from what I understand, that many members of the scientific community had a hard time accepting that revelation.

There were, indeed, a considerable number of scientists who never quite recovered from their initial discovery. But to understand their reaction, you must put it in the context of the time: it was nineteenth-century Christian Europe that had the brilliant idea of focusing on the origins of humanity, and the first discoveries were made in Europe itself, first in Belgium, then in Germany. And what a shock it was. The researchers had been convinced that they would come up with a presentable ancestor: after all, had man and woman not been created in the image of God? Imagine their surprise and chagrin when they came upon a stack of fossils of an individual who, they would later learn, hardly fit that preconceived godlike image.

Who was that person?

Neanderthal man, who was discovered in the Neander valley of Germany and dated back to the Paleolithic period. The problem was that this ancestor from the Stone Age was uncommonly ugly: he had a flattened skull and a puffy face, and the arch of his eyebrows was so overdeveloped that it virtually formed a visor. Many eminent scientists of the period did everything they could to malign this poor benighted creature. Some claimed that it was really no more than some hairy, arthritic creature. Others maintained that the only sound it could make was an "ugh!" Needless to say, it took many years before he was accepted into the human family, and even then only as a very distant cousin.

The Tom Thumb Technique

Whenever you "discover" an ancestor, it's actually no more than a few fragments of bone, a jawbone, or simply teeth. How is it possible to reconstruct accurately an entire skeleton with so few elements to work with?

The first remains discovered, which as you say are often teeth, are enough for us to figure out the person's entire morphology—the form and structure of a creature taken as a whole—and the alimentary significance of the rest of the body. We know, thanks to the laws of correlations of comparative anatomy, that a certain tooth fits into a particular jawbone, that this jawbone corresponds to a specific kind of skull, that this skull belongs to a certain type of appendicular skeleton, that this skeleton supports a particular kind of musculature, and so on. Starting with just the tooth, we can accurately reconstruct the entire animal by this deductive process.

And from that you can even go so far as to deduce the creature's development and behavior?

Absolutely. If, for example, you study the enamel of a tooth under an electron microscope, you can detect minute striae—little streaks or lines—invisible to the naked eye, that reveal the way the tooth developed and give us a pretty good idea about how the individual grew up. If we find the femur and see that its knee articulation is unstable, these observations translate into the knowledge that we're dealing with a tree-dwelling biped. But it goes without saying that the more elements we have at our disposal, the more precise the reconstruction.

Starting with the earliest research in the nineteenth century, and following all these little fragments of bone à la Tom Thumb, have scientists been able to trace the full evolution of man?

Curiously, we found fossils in the inverse order of their antiquity—modern men first, then their ancestors—which enabled us to recognize and accept them with a little less difficulty. But first we had to accept the idea that the origins of man go back much further than we ever believed.

Appearance with the Flowers

Is there a precise date that scientists agree on for the appearance of the earliest human?

We can't date with absolute precision the origins of mankind any more than we can the origins of life. Nor do scientists necessarily agree on a definition of "human." What we do agree on, however, is that human evolution has been going on for a very long time, and we agree on a zoological line of descent during which various characters made their appearance.

Do we at least know what the major stages were?

We have to go back to the end of the Cretaceous period, which means about seventy million years ago. It's the dawn of the third geological period; the last dinosaurs are dying out. The environment is undergoing profound changes, and we know that the history of evolution is intimately bound with that of climate. At the time, Africa was an island; so were South America and Asia. Europe, North America, and Greenland were all one continent, and it was on this continent that little animals began to appear: the first primates, which descended from insectivores, or insect-eating mammals. These primates began to proliferate in the midst of a brand-new flora: the world of flowering plants.

Born with the first flowers! What a lovely image.

It was also the period of the first fruits. These primates, which conquered this new environment, were in fact the first creatures to eat fruit. They broke away from the habits of their ancestors, who fed on insects. With the passing generations, this change would lead to a series of anatomical transformations. For example, the body would take on a clavicle, or collarbone, an innovation with all sorts of wonderful repercussions.

*

Such as?

The clavicle, which connects the breastbone with the shoulder blade, enlarged the animal's thoracic cage, therefore increasing the size of its upper limbs and enabling it to seize tree trunks and climb toward ripened fruit. For the same reason, the claws, which impeded climbing, would become flat fingernails. Among the primitive primates, the thumb was opposable to the other fingers, which allowed the animal's hand to seize fruit, a stone, or a piece of wood.

The Group from Purgatory

Who were these charming animals?

The oldest primate we know about was baptized *Purgatorius*, because the scientists who discovered it in the Rocky Mountains of North America were working under incredibly difficult conditions, in a real purgatory. It was no bigger than a rat. It lived in trees and fed on fruit, but it would not look down its nose at insects it they happened to come by.

And it was one of our ancestors?

Not directly. These little primates went on to colonize Eurasia, then the island formed by Africa and Arabia, which was covered with dense tropical forests. It was there that the real first ancestors, common to both man and the great apes, the higher primates, appeared about thirty-five million years ago. These large monkeys were isolated in Africa, which leads us to believe in a unique origin of the human lineage. It appears that about this time there was a great drought, which brought with it the selection and adaptation of new species.

Which ones?

In the Fayun Basin (the region where Cairo is located today) and in Oman dwelt a little four-legged monkey that was baptized *Egyptopithecus,* because it was first discovered in Egypt. It was about as big as a cat, had a long tail and a long snout, and stood apart from its predecessors by the fact that it had a slightly developed cerebral frontal bone, with a cranial capacity of forty cubic centimeters (as compared with fourteen hundred cubic centimeters for humans today). Although this capacity was very modest, it gave the animal the ability to react in a number of different ways. Thanks to the development of its nervous system, it was able to make use of new talents, new capacities. In particular, its visual capacity was greatly enhanced and superseded the sense of smell. The animal had stereoscopic vision, which, as a tree dweller, was a distinct advantage. At the same time, these prosimians made their first halting steps toward a social life: they communicated by mimicking one another.

*

How do you know that for a fact, since these animals have been extinct for millions of years?

Obviously, we haven't been able to deduce that from the direct observation of a little *Purgatorius,* but lemurs, which are alive and well and living in Africa, or tarsiers, which live in Asia, offer us very precise indications about their behavior, which is comparable in certain respects. Both lemurs and tarsiers have an active, well-developed social life. An examination of the cranial fossils of *Purgatorius,* and especially molds taken of the interior of their skulls, leads to the same conclusion. The dimensions of certain parts of their brain give us every reason to believe that they were already highly social.

Did they live together in families?

Elwyn Simons, the American scientist who discovered the species, told me that two skulls found at the same site showed an important sexual dimorphism—that is, the skulls were very different. One was in all probability a male skull, the other a female, which suggests that they did live in groups. All this leads to the conclusion that they were developing a form of communication and that they were quick minded. Simple, no?

A daring leap, in any event. What happened next?

Their descendant, the proconsul, lived in a forest farther south and had a more developed cranial capacity (150 cubic centimeters). There were, in reality, several species, the largest of which was about the size of a small chimpanzee today. The proconsuls had the good fortune to live at a time when major geographical changes were occurring: seventeen million years ago, the Africa-Asia plate came together with the Europe-Asia plate. The proconsuls and their descendants took advantage of this geological event to cross the "bridge" into Europe and Asia, where they flourished. Certain among them evolved and gave rise to a host of new species. The most noteworthy were the kenyapithecines in Kenya, but also the dryopithecines (the "monkey of the oak trees") in Europe and, somewhat later in Asia, the ramapithecines. For a while, we thought that this last-named species was part of our human family, but we were mistaken.

Fallen from the Tree

Until very recently, as I recall, *Ramapithecus* was depicted in those classic illustrations in our high school texts as the last in line, the one farthest to the left in the single file of our human ancestors. Now, as I understand it, he's been dethroned.

The biologists made us change our minds. Using the latest techniques to test the antibodies found on several fragments of teeth of a *Ramapithecus,* they discovered that they bore a close relationship not to the human race but to that of orangutans, which are found only in the jungles of Borneo and Sumatra.

"Orangutan," by the way, means in Malayan "man of the forest," so the Europeans who baptized it thought of it as a near-human "wild man." In any event, the same experiment with the antibodies on the teeth of *Australopithecus* showed that they were very close to humans. Biologists have also determined that man and the chimpanzee are genetically very close: 99 percent of our genes are common to both species. As confirmation, a *Ramapithecus* skull was discovered in Pakistan that was also morphologically very close to that of the orangutan. The case is therefore no longer in doubt: *Ramapithecus* is not our ancestor, but *is* the ancestor of orangutans.

With *Ramapithecus* fallen from our family tree, are we still searching for the "missing link" between man and monkey?

The expression "missing link" is inexact, because it presupposes that there is an intermediary between today's humans and monkeys. What we're looking for is the ancestor common to man and to the great African apes, the fork that separates the two branches, one leading toward the chimpanzees and the gorillas, the other to *Australopithecus* and, later on, to humans. It's all a question of date, of when that divergence took place.

And do we know when that was?

Biologists maintained that it occurred roughly five million years ago, whereas paleontologists generally agreed that it was much earlier, about fifteen million years ago. We compromised: we settled on seven million years. When we decided that

Ramapithecus was not a true ancestor, we moved the date of that great split forward and eliminated the orangutan from our ancestral tree. Since chimpanzees and humans are genetically very close, the logical conclusion is that we share a common ancestor. We thus abandoned the notion that human origins could be traced back to Asia and came to the conclusion that we had to look to Africa, to the descendants of the great monkeys in Africa, to find the ancestors of the human race.

The Primitive Plains

What ultimately made you focus on Africa?

The idea that Africa could be the cradle of the human race had already been put forth by Darwin, and later on by Teilhard de Chardin. After having worked throughout his life in Europe and then in Asia, the latter came back from a trip to Africa near the end of his life and proclaimed: "*There* is where we have to look for the origins of the human race; we're stupid not to have understood that earlier!" And then in 1959, when Louis Leakey discovered an entire skull in Tanzania, that intuition was confirmed. The age of the skull was calculated by measuring the natural disintegration of unstable isotopes, and the figure shocked the scientific community: the skull dated back 1.75 million years! At first, no one wanted to accept it.

*

The same old arrogance that prefers not to believe that the human race goes back that far?

Exactly. At that time, science had identified most of our human ancestors, but there was considerable lack of agreement about their status and how old they were (the first australopithecine had been discovered in 1924, but for a long time it was considered a relative of the chimpanzee). Scientists were convinced that the first human ancestor did not go back more than 800,000 years. But with the new methods of dating using radioisotopes and the extraordinary harvest of fossils discovered over the next decades, we had no choice but to push the date back.

So all eyes were now turned toward Africa.

Absolutely. Year in and year out, there are a number of international expeditions to Kenya, Tanzania, Ethiopia, to well-known sites: Lake Turkana, Olduvai Gorge, the Valley of Omo. I calculated that in all we have gathered no fewer than 250,000 fossils, 2,000 of which are either human or prehuman, most of them dating back two to three million years. This impressive crop has enabled us to reconstruct our genealogy.

Are we now sure that the human race began in Africa?

Science can never be "sure." But all the discoveries tend toward that conclusion. All we have to do is quickly run through the list of all the different places where fossils of human ancestors have been found. Fossils going back seven million years have been found in Kenya, where we've also found many that are five or six million years old. In Kenya, Tanzania, and Ethiopia, we've discovered fossils that are four million years old. In those places plus in South Africa and Chad, fossils three million years old have been uncovered. And in all those areas plus in Europe and Asia, two-million-year-old fossils have been found, together with a number of chipped stones. When we get to a mere million years, the sites extend throughout Africa, Asia, and Europe. Then we go on to Australia and the Americas. When you stack up all this information and date it chronologically, you can follow the long history of human population, and you are obliged to come to the following inescapable conclusion: man's origins were within a small area of Africa, and from there he spread throughout the African continent and then to the rest of the planet. And if you want to be fair and up-to-date, you can add: with a brief foray into the solar system.

The Elusive Grandfather

So we have unity of time and place: Africa, roughly seven million years ago. Do we now know the character who evolved on this primitive scene, our very first "grandfather"?

It's difficult to pin him down exactly. For the past twenty years or so, each time we find new fossils dating from that early period, we think that we've found the ancestor. *Sivapithecus, Kenyapithecus, Oranopithecus, Gigantopithecus, Oreopithecus, Otavopithecus*—all these various species have played this part at one time. The ancestor common to man and monkey is surely one of them.

But which one?

We don't know. If *Kenyapithecus* (fifteen million years old), discovered by Louis Leakey, isn't that common ancestor, then it's certainly one of its cousins. For one thing, its skull gives every indication that it was well adapted to life on the plains; for another, its eyeteeth were smaller, its molars larger, and the enamel thicker, and there was a differential use of the teeth in general, all pointing to the fact that the childhood of this hominid had grown longer.

Wait a minute! How can the enamel of the teeth provide information about the length of someone's childhood?

Simply because the different degrees of wear and tear on the enamel of successive teeth show that it took longer for the teeth to come in. And if teeth come in later, that means that it took longer for the species to become an adult, which means that the child spent a longer time with its mother. The proof: the teeth of this very ancient ancestor took three times longer to come in than did those of chimpanzees. The time a child spends with its mother is a time of learning, of apprenticeship. The longer the period of childhood, the more "educated" the species. And with *Kenyapithecus,* we detected an evolution of the kind I've been describing.

What do we know about this curious animal?

It was a large ape, a tree-dwelling quadruped, endowed with very solid upper limbs; and we know that it stood on its hind legs from time to time. Its brain was larger than that of its ancestors (three hundred cubic centimeters), its face was somewhat less prominent, and for a long time now, the species had not had a tail. It lived partly in the forest, partly on the plains. It fed on fruit, but also tubers and rhizomes—something we can determine from the tooth enamel, because you use your teeth more feeding on roots than on fruit. And we also know that it was a social animal.

The Benefits of Drought

What happened next?

Seven million years ago, this ancestor was living in the dense forests that covered Africa when a major geographical event occurred on Earth: the Great Rift Valley collapsed, and at the same time, some of its edges were raised, which over time formed a veritable wall. This geographical fault, formed by the valley's collapse, was enormous: it ran all across Africa eastward to the Red Sea, then up to the Jordan River, and ended at the Mediterranean Sea: six thousand kilometers (thirty-seven hundred miles) long and more than four thousand meters (roughly thirteen thousand feet) deep in Lake Tanganyika. An American astronaut told me that this great gash in the Earth is actually visible from the Moon.

And what were the consequences of that momentous event?

There was a radical change in climate. Rain continued to fall on the western slope, but on the eastern wall, the slope known as Ruwenzori, there was less and less rainfall. On that eastern side, paleobiologists tell us, the forest diminished, the flora changed. We can see similar phenomena in many parts of the world today, where it often rains on the western slopes of mountain chains but the eastern side is dry—such as in the western United States. The vegetation is very different.

So our ancestors were separated into two different populations.

Yes. Those that remained to the west of the divide continued to live their tree-dwelling life as before, but those isolated in the east had to confront the plains, then the steppes. This division into two environments resulted in two different evolutions over the course of countless generations: the populations to the west resulted in today's monkeys, gorillas, and chimpanzees; those to the east in the prehumans and then the humans.

*

On what do you base your hypotheses?

All two thousand human and prehuman remains that we've managed to gather over the years have been found on the eastern side of the Rift Valley. To be fair, to date, we haven't found any vestiges of premonkey ancestors in the western sector; if we had, it would obviously reinforce the theory. But even so, it's plausible. We're pretty sure that this tiny region of eastern Africa, which is shaped like a quarter of an orange, gave a new impetus to the evolution of primates that would lead to man.

Our cradle . . . In a sense, then, we were born out of drought.

Exactly. Everything that distinguishes us—the fact that we walk erect, that we are omnivorous, the development of our brain, the invention of our tools—is a result of adapting to a drier climate. It's the classic mechanism of natural selection: a small group of ancestors who genetically possess advantageous characteristics that allow them to survive better in this new environment becomes, as the generations go by, the majority of the population, since, by virtue of their living longer than the others, they tend to pass on to their descendants in ever-increasing numbers the same characteristics.

The Upright Monkey

What were those advantages?

We can only speculate. Perhaps a different growth of the pelvis, which gave them the ability to stand up more easily and therefore to see both prey and predator better, to attack and to defend themselves more efficiently, to carry food more quickly and easily to their children. Whether the upright position was the cause of that evolution or the consequence of it, we're not sure. In any event, we do know that those who had that advantage prevailed over the course of time. To save your skin in those early days, you had to have your wits about you.

What impelled them to assume once and for all the upright position?

Perhaps because of a genetic mutation, some individuals possessed a lower and larger pelvis, which meant that it was uncomfortable for them to go about on all fours. In this new environment, this so-called handicap actually turned into an advantage. And as generation succeeded generation, the upright position won out.

Is that a hypothesis?

Of course. Who can really know for certain? When we observe the behavior of chimpanzees, we see them assume the upright position in three situations: to see farther, to defend themselves or to launch an attack—since standing up frees their hands and allows them to throw stones—and to carry food to their offspring. We can imagine that our ancestors shed their body hair to facilitate perspiration, because of the dryness of the climate, and thus had to carry their babies. Before that, when their bodies were covered with hair, the babies did not have to be held; they simply clung to the mother's body hair. We can also surmise that the upright position, in these treeless plains, offered less of a target for the burning rays of the Sun and reduced perspiration.

Whatever the reason, we are certain that these ancestors adopted the upright position once and for all.

Yes. Study of the internal traces of the skull fossils we've found offers the same evidence: the convolutions of the brain are less noticeable on top than on the sides, which is logical, for if the body is upright, the top part of the brain is not pressing against the inside of the skull and therefore leaves fewer traces.

Scene 2

Our Ancestors Get Organized

Not yet men, no longer really monkeys, but standing up on their two hind legs, our early ancestors gaze out upon the world from their lofty position. They exchange words of love and eat snails.

The Australopithecines Hopping About

In eastern Africa eight million years ago, the prehumans were already well on their evolutionary way. They had broken away from the world of the great apes. How were they different from the species that had preceded them?

They were standing, and that's the way they would remain: upright. That was a real revolution. Their pelvis, their shorter upper limbs, their ribs, and even their skull were set differently on the spinal column. The entire morphology of their skeleton marks them as bipeds. Moreover, in Tanzania, researchers discovered their footprints in volcanic stone: they were the footprints of a 3.5 million-year-old biped. The English research team led by Mary Leakey noted that the footprints were crossed, as if the animal's gait had been unsteady.

And what did they conclude from that observation?

Perhaps, as one of the less serious scientists jokingly suggested, the consumption of alcohol went back much further than was previously thought. Perhaps the stone on which the footprints were found had slipped. Fortunately, we later found at the same site the footprints of an adult and a child, and they were perfectly normal.

So the honor of our early forebears was saved. How many species of *Australopithecus* were there?

For a long time, we thought that there was only one. As it turns out, their world was a lot more complicated. Starting eight million years ago, there was a real proliferation of species over the next seven million years. Among them were the antecedents of the human race, but that did not prevent other species from developing their own more classic descendants at the same time. So there are a number of different species that are roughly contemporaneous, and it isn't all that rare to find that an ancestor of one is also its cousin.

*

Is it possible to sort out all this proliferation?

Yes, of course it is! Everything naturally begins with the archaic species that are called *Motopithecus* and *Ardipithecus*. They don't go back any further than four million years. Then you have the australopithecines themselves, which pick up the baton in the evolutionary relay race from roughly four million to one million years ago. Bear in mind that all these species lived in eastern Africa, a vast territory divided into numerous depressions, which favored the diversification of species. We find, for example, traces of *Australopithecus,* which were baptized *anamensis,* in the region around Lake Turkana, which was fairly open; others of the same species were found in the more wooded areas in the Afar Basin, which were called *afarensis*.

Are we still discovering new species?

Yes, but our harvest is relatively modest, because the basins dating from four to eight million years ago, the essential period for understanding the appearance of Hominidae, are few and far between. Therefore, we have very few fossils, but even if we're not completely sure how these species derived from one another, we have enough evidence to figure out the major affiliations.

What did prehumans look like?

The fossils that have been most closely examined are, as you know, the remains of "Lucy," a young female who lived three million years ago. Her skeleton is the most complete—or, more accurately, the least incomplete—of any that has ever been discovered.

Lucy's Knee

This is slightly off the track, but I've always wanted to ask you, since you were a member of the team that discovered Lucy, if it's true she was named after the Beatles' song.

Absolutely true. When we found her in 1974 in the Ethiopian Afar, we had a number of cassettes with us, including one that contained the Beatles' song "Lucy in the Sky with Diamonds." Since the find was a young female, the name stuck immediately. The Ethiopians preferred to baptize her *Birkinesh,* which means "person of value."

She certainly was a person of immense value in the search for human origins, not only for her worldwide notoriety but also because of what we learned from her, isn't that true?

Absolutely. She has been studied piece by piece. Numerous theses have been devoted to her arm, her neck, her shoulder blade, her knee. Much is still unclear. We can't even be entirely sure that she was a female, as she has some male characteristics.

Could you describe her for us?

She was less than three feet tall. She was slightly round-shouldered, with her upper limbs slightly longer than ours in relation to the lower limbs, a small head, and hands capable of grabbing objects but also of taking hold of branches. She was a biped, but she was also a tree climber.

So she walked the way we do.

Not quite. When we compare the different ways various individuals walk—adults, children, contemporary chimpanzees—we can deduce fairly accurately how walking has evolved over the course of time. Lucy doubtless took shorter steps than we do, and her pace was more rapid, a kind of rolling gait, not unlike trotting. We even succeeded in reconstructing what childbirth would have been like for Lucy, by measuring her pelvic area and deducing from it the probable size of the fetus. From all we can tell, the movement of Lucy's babies at birth—assuming she had children—was very much like that of human newborns today, as opposed to that of baby monkeys.

What else do we know about Lucy?

Although she was two-legged, she was still very much a tree climber, as I mentioned earlier, as is proved by the articulation of her various joints. Her elbow and her shoulder are more solidly fitted than are ours, which provided security as she moved from one branch to another; her toes are long and curved, and her knee possesses a considerable amplitude of movement—aptitudes that are consistent with one who climbs trees and leaps from branch to branch. She was a social animal. Like all primates, she was a vegetarian. The thickness of the enamel of her teeth shows that she ate fruit but also roots. And from the amount of their wear and tear, we think that she died before she was twenty, either by drowning or having been eaten by a crocodile, since we discovered her in sediments of what had been a lake.

Poor ancestor . . .

Don't shed any tears for her. In all likelihood, she's not our ancestor, but rather a branch that derived from it, because her characteristics are archaic. During that same period—about 3.8 million years ago—the *Australopithecus anamensis* or *africanus,* from South Africa, was endowed with a knee that was more human than Lucy's. The prehuman species perhaps evolved simultaneously. And because two species have comparable characteristics doesn't necessarily mean that they belong on the same family tree. Compare fish and marine mammals: they resemble each other, yet they're totally different. The ancestors of marine mammals are terrestrial quadrupeds that at some point decided to return to the water.

With Their Hands Free

So we don't know our true australopithecine ancestor.

No. Personally, I have a slight preference for the *anamensis*. It's the right period—four million years ago—and the morphology of both its upper and lower limbs is practically modern, which makes it a biped similar to us today. Lucy, on the contrary, still harbors several tree-dweller characteristics. Then there are more hardy *Australopithecus* species that arrive on the scene.

What do they have that the other species don't?

They're considerably steadier on their feet, and they're better walkers than their predecessors. The brain is still modest—about five hundred cubic centimeters—but it's better irrigated. The structure of their teeth is different, which enables them not only to chew but also to grind, because with fewer trees and therefore less fruit, their diet has become rougher, more fibrous. In addition to the remains of australopithecines more than three million years old, the various fossils we've found in the Omo Valley of Ethiopia have produced a considerable number of cut stones.

You mean to say that *Australopithecus* was already using tools that far back?

*

We still have a hard time accepting that notion, but apparently they were the first people to use tools. The traces on the stones that have been found indicate that they were used to peel roots and tubers, not to cut meat or scrape bones. It's entirely possible that such stone tools were used by the australopithecines of Lucy's family. What this means is that tools were being made by people whose hands were not yet completely free.

The Brain as Tenant

Some anthropologists have proposed the following seductive scenario: having discovered the tool, the prehuman needed both to free his hands and to assume an upright position. In other words, they have reversed the cause and the effect. As a result, the size of the braincase was able to increase, and the brain as well.

That's completely probable. Fish had no problem supporting their heads, because the head was one with the rest of the body. But as soon as it began to develop lungs and drag itself along the earth, the terrestrial quadruped had problems holding up its head, which was becoming more and more independent. And when the prehuman began to stand on two feet, the problem became greater. The upright position liberates the head and at the same time allows for the braincase to increase in size. The brain, as would any worthy tenant, follows suit and expands to fill the available space.

And from that point on, it can acquire new capabilities, new aptitudes?

It's also possible that the increase in the size of the brain brought about another change: a shortening of the term of pregnancy. Since the brain of the fetus is larger than that of its predecessors, it follows that delivery had to occur earlier. And that, in turn, allowed for the further development of the brain after birth. It also appears that the birth of children headfirst, as opposed to the breech position, was an effect of the upright position. Another obvious advantage: *Australopithecus,* because it stands up, uses its hands more often and has a greater and greater ability to perfect its tools.

And yet tools are also being used by monkeys.

True. The use of tools was not limited to humans or pre-humans. Monkeys, for example, were extremely adept at stripping the leaves from the branches of trees in search of termites or at using stones to crush the shells of nuts. But to make one tool by using another was apparently a superior stage that monkeys never attained.

How did the australopithecines of that era communicate with one another?

*

It's probable that they had all sorts of things to say to one another, but their means of doing so was limited to mimicking, making signs, or voice inflections, for at this stage of evolution they did not have the mechanical possibility of articulate speech. Look at the chimpanzee: for a long time we tried to teach chimpanzees to articulate, to say at least a few words, until we realized the problem—their palate wasn't deep enough, and the way their larynx was positioned made speech as we know it impossible. When we then had the bright idea of trying to teach them sign language, we noted that they not only registered several hundred concepts but also were able to make associations. What is certain is that language came into general use with that other individual who appeared roughly three million years ago, a creature who was taller, straighter, less a climber than the prehumans, a creature endowed with a more developed brain and vascular system: man.

An Opportunistic Individual

Did *Australopithecus* cohabit with humans?

For a million years at least, maybe two. They probably didn't occupy the same territory, but they did cross paths once in a while.

*

And, I assume, became rivals?

Why would you assume that? I know that people love to decorate the distant past with all sorts of dramatic images. Just think of the images of prehistory we've all seen, in which our poor beleaguered ancestors seem to be living in a state of constant terror: volcanoes erupting in the background, fires sweeping the forests or plains as early man flees from a horrible wild animal or from the clutches of a huge australopithecine armed with a club. Or, conversely, pictures depicting early man, suddenly quite civilized, lying in wait to launch an attack on some frightful, hairy monster.

Both of which, I gather, are not accurate portrayals?

Not quite. It's true that the human brain was at this point capable of developing strategies and planning concerted acts against australopithecines. There were doubtless battles between them, but they weren't what I would call well ordered. They were relatively few, in fact; in the main, the two populations coexisted. If you doubt what I say, look at the Masai, a people who until recently were living much as primitive humans did in the period I'm talking about: they manage to walk among lions, rhinoceroses, buffaloes, all sorts of wild animals without the slightest incident, which goes to show that one can live in peace with one's fellow creatures, that is, in harmony with one's environment—bearing in mind, of course, that you have to keep your eyes peeled whenever you enter another creature's domain. Having said that, I'm willing to admit that

that equilibrium can easily be broken and that one of the gentle Masai who is less alert than he should be might well find himself serving as another creature's evening meal. So, getting back to the cohabitation of humans and australopithecines, I think we can assume that despite all the above there were times when a human hunted down and ate an australopithecine child, which was probably delicious and so much more tender than an adult.

Are you serious?

Absolutely. The early humans were omnivorous. Any "game" that happened by was fair game. This said, I'm not suggesting that the disappearance of *Australopithecus* was a result of some kind of mass extermination.

Then how did they disappear?

By the classic mechanisms of natural selection. About a million years ago, in an environment that was steadily becoming drier and cooler, *Australopithecus* was less and less able to adapt and therefore was becoming more and more vulnerable.

Were they in competition with humans?

Yes, but that does not imply that there was any violence between them. Flat oysters disappeared as a result of the pressures on them by the so-called Portuguese oysters, and so far as we know, there weren't any pitched battles between the two. The

Portuguese oysters simply adapted wonderfully well to the domain of the flat oysters and proliferated there.

Australopithecines, in a sense, were too close to humans.

Yes. And contrary to humans, they were unable to transcend their ecological niche and remained inferior to the environment around them. As a result, their species became less and less fertile, and after several hundred thousand years they became extinct. Humans prevailed: they were tall, more erect; they were omnivores; they were extremely opportunistic and, increasingly, outfitted with better and better tools.

A Spate of Human Species

Three million years ago, in the landscape of ancient prehumans, there was the *Australopithecus* species going about its business on its hind legs, but also the very early representatives of the human race, which were beginning to hunt for their supper. Did that make the landscape a trifle crowded?

Fairly. Two universes came together: that of the prehumans, who were going to disappear from the Earth, and that of humans, who had just been born. We tend to classify the latter into three different categories: *habilis, erectus,* and *sapiens*. But we've recently discovered other subspecies, such as *Homo rudolfensis* and *Homo ergaster.*

Why are there so many species?

Because the descendants of *Australopithecus* were so abundant. It's difficult to establish clear lines among all the various pre-human populations. The *Homo* species, however, evolved in such a regular manner that, as far as I'm concerned, *habilis, erectus,* and *sapiens* are simply different stages of the same species.

In other words, we should refer to "man" in the singular?

Yes. The human race.

What distinguishes that race from the others?

Its feet. One of mankind's most recent acquisitions is a very special foot that is the result of walking on two legs, with the toes parallel. Although his upper limbs are less solid than those of his ancestors, his lower limbs are more stable, for he is less prone to climb trees. His jaw is more rounded, with the canines and incisors more developed relative to the molars, which are smaller than those of *Australopithecus* because of his omnivorous eating habits. And finally, his brain is both considerably larger and endowed with complex convolutions.

*

At this early stage, is the human species hairy?

Probably not any longer.

Black?

We have no way of knowing. But we can deduce that he is certainly dark skinned, given the hot, sunny climate and the relatively treeless landscape. About two and a half million years ago, we know from studying the flora and fauna that there was a major climatic crisis: a period of great drought.

Did it have an evolutionary impact comparable to that of the rift divide that created *Australopithecus*?

Yes, and it resulted in enormous upheavals. Both the flora and the fauna were modified. The trees disappeared and gave way to grassy plains; a great number of animal species died out. *Australopithecus,* big in body but small in brain, used its powerful jaw to feed on tough, fibrous plants, tubers, and hard-shelled fruit. Man, with his larger brain and longer, slimmer molars, fed on a mixed diet, both meat and plants. In fact, both the australopithecines and the human species are doubtless the products of selection caused by that profound change in climate.

The Drought of Love

What did the omnivores feed on?

Fruit, grain, tuberous plants. Animals big and small, from elephants, hippopotamuses, and gazelles to chameleons and frogs. The bones of their meals demonstrate that they had an extremely varied menu. Because they had such strong teeth, they were able to crack open both grains and hard-shelled fruit. As we can tell from the skulls of some animals that died as a result of being hit by stones, man at this stage was a professed hunter. Then, as now, man was a creature who ate anything and everything that came his way—a most opportunistic fellow, as I said before.

We also know that once he found food, he brought it back to his fellow creatures, which is a major milestone. Monkeys either eat their prey themselves or try to steal it from other monkeys. Not so for humans: for the first time ever, here's a creature who shares, which means there is a form of social organization. And about two million years ago, man also began to try his hand at creating primitive shelters, protections that were either round or in the form of half circles. A number of remains of these shelters have been discovered and date back to this period.

How did these early humans communicate?

*

Man's adaptation to the new conditions of climate resulted in a modification of his respiratory system and a lowering of the larynx. Man is the only vertebrate whose larynx is in a low position. The installation of a kind of sound box between the vocal cords and the mouth, combined with the deepening and reduction of the lower jawbone behind the incisors, gave the tongue greater mobility. Even if it wasn't articulated the way ours is, language became more and more developed. Certain studies of the skulls of very early man reveal the presence of a region of the frontal lobe that today corresponds to the seat of language, the so-called Broca sector.

And all that was the result of the change in climate?

Evolution does in fact depend on outside events, and often that event is environmental. Still, one would be hard-pressed to claim that man's larynx descended solely so that he could start speaking.

But what you are saying is that not only man's body but also his language and culture were the result of this prolonged drought.

It's a good explanation, in any case.

What about love?

You're going to say that I'm exaggerating, but for me, love is also the fruit of this prolonged dry spell. The drought, and with it the more exposed conditions of the world that early humans lived in, logically brought them closer together, as it reduced the period of pregnancy. And that, in turn, meant that mother and child necessarily spent more time together after the child was born. This, helped along by the appearance of consciousness, gave birth to emotion. And perhaps during this same period, the father was obliged to spend time with the mother-child couple, at least for the mating period. Feelings between men and women may have come into being at the same time. Edgar Morin said to me one day: "Freud was trying his best to get rid of the father, and you prehistorians are bringing him back into the picture to explain the flowering of humanity." And to some degree, he's right.

Scene 3

The Human Conquest

The old world was dying; a new world was born, dominated by an opportunistic biped who conquered the planet. He invented art, love, and war and began to ask questions about his origins.

The Spirit of the Hill

The early representatives of the human race are garrulous and loving. Within a relatively brief time, they are going to set about colonizing the earth. Is it because they are innately curious?

Why would they wait hundreds of thousands of years in the same place they were born before checking out what lay beyond the next hill? When you climb the nearest mountain to see what lies on the other side and you see that in the distance lies another mountain, you want to climb that one as well. And besides, early man possesses a degree of intelligence. He needs to hunt to find food, and that obliges him to travel. He has what it takes to dominate: I imagine that he must have been rather imposing as he stalked his prey and threw his stones at them.

Did early man live as a family?

Probably in small groups of about twenty to thirty people. We've noted similar communities among the Inuit hunters of Greenland. When the population increases and reaches a certain threshold beyond which the group can't function properly, it splits into a new group for reasons of survival. The new group detaches itself from the original one and goes off in search of food elsewhere, usually a few kilometers away. In the early years of the human race, the population increased considerably.

How can you know that for a fact?

In any given environment, there is a relation among the numbers of plant eaters, meat eaters, and omnivores. By calculating the proportion of human fossils discovered in an archaeological site of a specific historical period, we can estimate the total population whenever the numbers are sufficient to be statistically accurate. What we found was one human per every ten square kilometers of space, which is exactly the density of the aboriginal population in certain regions of Australia.

So early man, starting with small clusters of people, began to colonize the planet.

Yes. If we assume for each generation a movement of only fifty kilometers, or about thirty miles, that would have been enough to bring man from his African origins to Europe in no more than fifteen thousand years, which in relation to our overall historical time frame is a mere blink of an eye. Or, to put it another way: fifteen thousand years is not even the margin of error of our various datings. Starting from the cradle of his African birth, man would fan out both west and east, into what is now the Americas and the Far East. In both regions, we have found fossils and cut stones going back more than two million years.

Well-Worked Flint

Are we still talking about the same humans?

At first, we're talking about two categories of early man, *Homo habilis* and *Homo rudolfensis,* then later on about *Homo ergaster* and *Homo erectus*. But since we have found intermediary fossils, it would seem that, following an explosion of East African forms, the conqueror of the world was one and the same species of man, which has been labeled with names indicating the various stages of evolution: *habilis, erectus, sapiens.*

What were the basic characteristics of *Homo erectus*?

First, he had a larger brain (nine hundred cubic centimeters) than that of his predecessor; his behavior was in general more sophisticated, both in the way he fashioned his tools and in the way he occupied his territory. Now, instead of making tools by striking stone against stone, he used either a sharp piece of wood or the horn of an animal to work the stone, which enabled him to control the chipping process more closely and thus produce more delicate tools.

A million years chipping away at flint stones! That's how long it took to find the right cutting edge.

Human progress was indeed slow. Prehistory, a number of historians believe, can be interpreted by the study of these knife edges. By comparing an equal number of worked flint stones from each of the major historical periods, they noted that fashioning cutting edges became a more and more efficient process. Three million years ago, a total of ten centimeters of cutting edge was obtained from a kilo of stones. With the introduction of double-edged stones, the total yield increased to forty centimeters of cutting edge, and later, to two meters for the tools of Neanderthal man (fifty thousand years ago) and to twenty meters for those of Cro-Magnon (twenty thousand years ago). The more we move forward in time, the more precise the stone cutting.

In what way more precise?

There was, for example, a certain type of stone cutting, known as the Levallois technique, that required a dozen very precise blows for the stone to splinter in a specific way, which presupposed the ability to plan and a considerable degree of abstraction. One prehistorian compared the technique to that of making an origami bird: you fold the paper in a certain way—once, twice, fourteen times—and if you've done it right, you can move the bird's tail up and down. But that takes real knowledge and sophistication.

Disorder at Home

Still, you have to admit that despite the development of the brain, progress was snail-like.

Quite true. *Homo erectus* went about armed with his two-faced tool for hundreds of thousands of years. Modern tools and weapons, from the earliest use of metal to nuclear weapons, were invented in a lightning flash compared to that. When we studied various archaeological sites in East Africa, we noted that a major change occurred about 100,000 years ago. From that point on, it seems that cultural changes superseded anatomical modifications. Evolution finds new responses to the demands of the environment. Achievements take over.

And did that bring about a change in the social organization of humans?

When you study the sites of early human habitation, those of *Homo habilis,* you find a real mess: everything is all mixed together—the remains of food, of cutting tools, of tools that were used to cut the meat. It appears that everything was taking place on the same spot. Later on, with *Homo erectus,* you find various dedicated areas of the encampment for different functions: there is one place where people slept, another where they ate, another where they cut their stones to make tools. That, of course, suggests a form of organization. And even later on, these various segregated places were spread farther apart, sometimes as much as several hundred meters. But you also find a central spot, a hearth.

Was it *Homo erectus* who invented fire?

Yes, about 500,000 years ago. Fire might well have been mastered earlier, but society wasn't ready for it. It's not by chance that fire came into use at the same time as the flint and when man had learned to work stones using the Levallois technique.

It's entirely possible that a number of geniuses had discovered better or cleverer ways to cut stone, but all societies tend to turn their backs on their inventors if they're not predisposed to understand them. You have to wait until the majority is sufficiently mature for any new idea to be accepted and put to use.

The Man with the Visor

And about now *Homo erectus* disappears from the scene, yielding its place to *Homo sapiens,* modern man.

The latter derived from the former, but very slowly, by a long process of evolution. It was a gradual transformation, which was taking place in the same way almost everywhere in Asia and Africa. The exception was in Europe: Neanderthal man.

The fellow who sent such a collective shudder of disgust through the entire body of early researchers. What was he all about?

Neanderthal man descended from *Homo habilis,* who populated Europe about two and a half million years ago. As a result of the successive glaciations that began in the Pliocene period in Europe and lasted to the Recent epoch, Europe became a kind of island cut off by the Alps and by the northern regions that were covered with ice. The early *habilis* found themselves isolated, in the strict sense of the term, and did not evolve the way their brethren did on other continents.

Why was that?

We know that on any island, over time the flora and fauna begin to evolve differently from those of the neighboring continent: they undergo a genetic drift. The older the island, the more diversified its flora and fauna, and the greater the differences between its flora and fauna and those of the continent. In the same way, if you isolated a group of human men and women on another planet, after a period of time, that planet's population would become different from us. Neanderthal was thus born of a similar genetic drift. He had no forehead to speak of, a suborbital visor, a puffy face.

None of which worked to his advantage, apparently.

Nonetheless, he lived in Europe from about two and a half million years ago until thirty-five thousand years ago and managed to cohabit with another *sapiens,* Cro-Magnon, who was called that because his remains were discovered in a cave by the same name near Dordogne in France. Cro-Magnon, unlike Neanderthal, had evolved in Asia and Africa and arrived in Europe relatively late, about forty thousand years ago.

The First Cohabitation

How did the cohabitation go? It's hard to imagine them going to war against each other.

For a long time they were described as opposing types: Neanderthal was thought to be barbaric; Cro-Magnon, civilized. As a matter of fact, they were close. They dwelled in the same sites, one after the other. Their tools were very similar, as were their lifestyles. Neanderthal was clever, creative. He possessed a well-developed language. He buried his dead. He gathered various objects for no other reason than that he liked them; we have collections of fossils and minerals from Neanderthal dwellings going back eighty thousand years. He also adapted very nicely to the technological changes taking place in the Upper Paleolithic period: the lamellate industries of the Charente-Maritimes and Yonne areas of France once attributed to Cro-Magnon were actually the fruit of Neanderthal man.

Did the two populations ever integrate in any way?

We don't know. We've never found any fossils that have the features of both. That's why certain researchers believe that we're dealing not with the same species but with two different species.

*

But Neanderthal did finally disappear. Why? One is tempted to suggest that he was done in by Cro-Magnon.

There's a grotto in southwestern France in which we've discovered a Neanderthal layer, then a Cro-Magnon layer, then another Neanderthal, then another Cro-Magnon, as if there were successive occupations, either seasonal or as a result of various aggressions by one or the other. Were there any battles? I'm of the opinion that Neanderthal simply went quietly into the night. Cro-Magnon was better equipped, both culturally and biologically, than Neanderthal. If there was competition, perhaps it wasn't violent. In any case, there is no doubt that Cro-Magnon prevailed, and Neanderthal became extinct.

Art and Manner

When you say "Cro-Magnon," you mean you and me, right?

Yes, modern man. He had a slender skeleton and a well-developed brain than enabled him to further develop his symbolic thought. He would go on to colonize the entire planet, fanning out in every direction, passing over the Bering Strait that had emerged 100,000 years before Christopher Columbus.

And, taking to rafts, he crossed the oceans and landed in Australia at least sixty thousand years ago.

And in Europe, Cro-Magnon and his descendants were here to stay.

In Europe, these descendants in particular would indeed do something that early man had never done in either Africa or Asia: starting forty thousand years ago, they would leave visible proof of their imaginative powers by drawing objects on the walls of caves.

The oldest caves in which these drawings have been found go back some forty thousand years. Do you see this as the origin of art?

No, the origin of art was a progressive achievement. Despite the anatomical discontinuity between Neanderthal and Cro-Magnon, there was a real cultural continuity. Neanderthal men were unusually curious. They gathered minerals, they made holes in shells and teeth to make necklaces, they fashioned musical instruments—whistles, little flutes—out of bones. The use of ochre goes back even further, several hundred thousand years, in fact.

To bury one's dead, to paint, to do something simply for the pleasure of doing it, to resort to rituals—all that implies the notion of time, an awareness of being part of a universe.

*

Consciousness and its consequence—symbolic thought—have developed slowly over many generations. But what is new, beginning roughly 100,000 years ago, is man's ability to conceive of another world to such a degree that he prepares for the voyage, with rituals to go with it and, starting forty thousand years ago, the art to go with it. Only certain special individuals have the right to this ritual burial, which implies a social pecking order.

The Relay of Culture

And then followed bronze, iron, then writing—history as we understand it today. Not to mention war. I'm right in assuming that war is one of modern man's inventions, am I not?

Yes, but it's a recent invention. The first charnel houses that we've discovered date from only four thousand years ago, the age of metal. It's as if the discovery of agriculture and animal breeding, followed by the discoveries of copper, tin, and iron, brought the notion of property, and therefore the need to defend your worldly goods and possessions. It's true that the production of metals implied the ownership of mineral deposits. And those segments of the population who found and worked them unexpectedly became wealthy.

*

As culture flowered, man exercised increasing control over his nature. Did his body continue to evolve, from early Cro-Magnon man until today?

Very slightly. As I said, his skeleton became somewhat more slender, as did his muscular system; man's teeth became shorter and fewer in number. As for the time of pregnancy, it became shorter. Mother and child grew closer to each other; the period of learning, of apprenticeship, grew longer. And the population was increasing by leaps and bounds: three million years ago, the population was 150,000 humans, all living in a small corner of Africa; two million years ago, there were several million humans; ten thousand years ago, there were ten to twenty million; two hundred years ago, a billion; and today, roughly six billion.

And later on, the human race became more and more diversified. Does the concept of race mean anything to you?

Not really. In botanical or zoological terms, a race is a subspecies. In the case of the human race, the concept of race has been much abused: we are all *sapiens sapiens*. To be sure, there are populations in which some individuals are closer to one another than they are to the rest of the population, but there are no such things as human *races*. We're all one race. The differences that exist are at the level of tissues, cells, molecules—distinctions that are meaningless.

Eve and the Apple

In the scenario of human evolution that we've just run through, what mysteries still remain, in your view?

The chief mystery is just how evolution works. In a changing environment, animals and man are capable of changing in order to adapt to new conditions of climate, as if there were, on each occasion, a pattern of mutations sufficient to ensure the correct choice. Evolution proceeds by natural selection, no question. But is that enough to explain the wonderful adaptability of humans to the changes in their environment? Or is it the environment that directly induces these genetic changes? Perhaps someday we'll discover the answer to that mystery.

Would you say that our history makes sense, that it has a logic to it?

All I can do is record what I know: living creatures today are more complex than those that existed a billion years ago. And personally, I believe in neither chance nor contingency, both of which seem to appear only when you study short periods of time.

Could that mean that we should reconcile the scientific conception of our origins with that of religion?

*

That's not out of the question. Science, after all, only observes. It cannot be dogmatic. Science knows that reality is always more complex.

Where would you fit Adam and Eve into our story?

Again, speaking only for myself, I suspect that they are both *Homo habilis,* living in the lovely scented plains of East Africa some three million years ago, not far from the vast geological fault I mentioned earlier. That region must have been a kind of terrestrial paradise when man began hunting and speaking.

A paradise replete with serpents and apples?

There was no scarcity of snakes, and as for apples, they were in all likelihood coconuts, the fruit of the palm tree. But one shouldn't try to stick science and the Scriptures together; it just wouldn't make any sense.

With Heavy Heart

What makes the human race so special? What differentiates it from all other creatures large and small?

It's more a question of degree than of nature. When we observe chimpanzees, we're struck by their resemblance to us, by certain humanlike behavior: male chimpanzees, for example, dance in front of females at the first rain of the season. Lévi-Strauss founded his perception of human society on the taboo of incest between mother and child. That taboo, you might know, is also operative among chimpanzees.

So how do we define the human race? Is it our self-awareness? Is it the capacity to love?

Certainly emotion is one of the elements that define us. But I would say, above all, our defining feature is the awareness of death, which operates on a whole other, higher plane. To realize that each of us is unique and irreplaceable, that the disappearance of a single human is an irretrievable drama, for me is the basic definition of reflective consciousness. That notion includes, of course, an awareness of self, of others, and of the environment, and also the notion of time.

What do you see as the moral of this long story?

What this last act teaches us is that, first of all, our origin is unique. We are all Africans originally, born some three million years ago, and that should incite us to think and act fraternally. We should also bear in mind that man emerged from the animal family after a long struggle against nature, by imposing his culture against innate determinism. Today we are wonderfully free—we can experiment with our genes, we can produce test-tube babies—but we should also be aware of how vulnerable we are. If one of our offspring were kept isolated from the modern world, if he or she grew up alone in the wilderness, that child would be totally deprived, unable to learn: the child would not even be able to walk on his or her hind legs. It took the full course of evolution, of the universe, of Earth, of humanity, to acquire this fragile freedom with which we're blessed, the freedom that gives us our dignity and responsibility. And if today we ponder our origins, be they cosmic, terrestrial, animal, or human, it's so that we can see ourselves and our world in a better perspective.

Epilogue

> Huddled on their little planet Earth, threatened by powerful weapons that they themselves have wrought, these conscious, curious creatures lift their eyes to the heavens and ask themselves anxiously: what will be the next chapter of this most extraordinary history?

DS: So here we are, after fifteen billion years of evolution and several thousand years of civilization. Is evolution, which has been taking place ever since the Big Bang, which has constantly led to increasingly complex structures, of which the human race is the crowning achievement to date, still going on?

JDR: The particles, atoms, molecules, macromolecules, cells; the first organisms made up of several cells; the populations consisting of several organisms; the ecosystems made up of various populations; and then man, who exteriorizes his biology—of course evolution is still going on. But now it's above all technical and social. Culture has taken over.

*

In which case, we would be at a new turning point of history, a rupture comparable to the appearance of life.

Yes. After the cosmic, chemical, biological phases, the curtain is going up on the fourth act, an act covering the next thousand years in which humanity will play the starring role. We are on the verge of attaining a self-awareness that is becoming collective.

How would you characterize this next act?

We might say that we're in the process of inventing a new form of life: a planetary macro-organism that encompasses the living world and all human production. This macro-organism is also evolving, and we are the cells. This world possesses its own nervous system, of which the Internet is only an embryo, and a metabolism that recycles various materials. This global brain, made up of interdependent systems, links mankind with the speed of the electron and is turning our exchanges upside down.

And if we keep the metaphor, can we speak of a selection process that is no longer natural but now cultural?

I think so. Our inventions are the equivalent of mutations. This technical and social revolution is moving forward far more quickly than did the Darwinian biological evolution. Mankind is creating new "species": the telephone, the television, the automobile, computers, satellites.

So humans are now making the selections.

Yes. What is the market, for example, if it's not a Darwinian system that picks and chooses, eliminating some inventions while promoting others? The major difference between this new evolution and that of Darwin is that man can invent in the abstract as many new "species" as he wants: this new evolution is being dematerialized. Between the real and the imaginary worlds, it is introducing the virtual, which enables it not only to explore artificial universes but also to test and manufacture objects or machines that have not hitherto existed. In a certain way, this cultural and technical evolution follows the same "logic" natural selection did.

Can we therefore say that complexity continues unabated?

Absolutely. But little by little, it is freeing itself of the mantle of matter. In a certain sense, we're coming full circle, back to the Big Bang. The explosion of energy fourteen or fifteen billion years ago looks very much like the opposite of what Teilhard de Chardin referred to as the "Omega Point," by which he meant an implosion of the mind freed of matter. Putting the element of time aside, it would be very easy to confuse the two.

*

It's difficult, though, to leave out the element of time and the very brief life span allotted to us humans. Does the individual still have a future if he is destined to become integrated into a planetary agglomeration that transcends him?

Of course mankind has a future! I'm of the firm opinion that the perfectibility of man still lies ahead of us. When cells come together, they attain a greater individuality than when they remain isolated. It's true that the stage of macro-organization involves the risk of planetary homogenization, but it also contains the germs of diversification. The more the planet becomes global, the greater the differences.

You describe today's society in biological terms, talking about its evolution, its brain, about mutations or their equivalent. Are you sure you're not taking your metaphors for reality?

There's no way that we can portray society in solely biological terms. If we were to do so, it would lead to unacceptable ideologies. This said, biology can "irrigate" our thoughts and reflections. In the early years of the twentieth century, mechanical metaphors—gears and clocks—dominated our thinking. Now the metaphors of our pedagogues tend to be those of an organism, which is fine as long as you don't take them literally. The planetary organism that we are creating exteriorizes our functions and our senses: our vision via television, our memory via computers, our legs via various means of transport. The major question remains to be answered: are we going to live in symbiosis with this planetary organism, or are we going to become parasites and destroy the host to which we're attached, which would doubtless lead to serious economic, ecological, and social crises.

And which do you think will happen?

At present, we're in the process of using for our benefit our sources of energy, of information, and of material, and we spew back out into the environment the waste products therefrom, each time impoverishing the system that supports us. We are making parasites of ourselves, since certain industrial enterprises are curbing the growth of others. If we continue along this path, we will indeed become parasites of the Earth.

What do we have to do to avoid that? To preserve the planet?

It's not a question of sequestering all manner of living things in some kind of enclosure, as some well-meaning ecologists would like to do, but rather of finding the harmony between the Earth and technology, between ecology and the economy. In order to avoid crises, we should draw on the lessons about the evolution of complexity found in the present history. Understanding our history can provide us with the necessary perspective, a direction, a "meaning" to what we're doing, and doubtless greater wisdom as well. Personally, I believe in the growth of collective intelligence, as I believe in technological humanism. And I live in the firm hope that, if we so desire, we can approach the next stage of humanity with a feeling of serenity.

The Future of Mankind

According to Professor de Rosnay, our history of the world is about to begin a fourth act, the title of which will be "The Cultural Evolution." Do you agree with that assessment?

YC: One day I said to an explorer who had just returned from the North Pole: "I can only imagine how cold you must have been up there." To which he shook his head and said: "I wasn't cold at all. I was always warmly dressed." That's fairly typical of our cultural evolution. We are constantly improving the mastery of our body and of our environment, and in this relay race of evolution, we have indeed passed the baton on to culture. From this point on, culture rather than nature will respond to the changing demands of the environment.

It's your opinion, therefore, that physically, *Homo sapiens'* body will not change anymore.

It will change, but only very slowly. You have to view it in very long terms, far beyond the next millennium. In ten million years, for instance, I'm quite sure that our heads will be different from those we have today and that our brains will continue to develop. Our skeletons will also become more slender.

*

Which means that the species will potentially be endowed with new aptitudes.

Certainly. For one thing, if the size of the brain does increase, it follows that the fetus's head will also be larger, which, as we've seen throughout the history of evolution, results in a shorter term of pregnancy. If the mother of this future superhuman gives birth, let us say, in six months instead of nine, the duration of childhood will be lengthened, and the time of early learning as well. We're not sure just how long pregnancies were in the early days of human history, but it's reasonable to suggest that as time went on and the size of our brain increased, the term grew shorter, as it will in the future.

So our biological evolution isn't really over yet.

It has slowed down, but it's still going on. We are still subject to the laws of biology and to its adaptations. For one thing, viruses are still evolving and can cause us problems in the future. Nor is there any way we can protect ourselves from some cosmic cataclysm that would modify the Earth's atmosphere. We can no longer say that man is subject to real natural selection.

No further major mutations of our genes that might change our species again?

Of course there could be mutations. But the homozygotes necessary to cause the mutations are another matter altogether.

In the present human population, the gene pool is permanent. There are no longer any isolated groups that might be capable, through some genetic shift of recessive characteristics, of any meaningful mutation of the species. But there is one notable exception: if, as is likely, man ends up establishing colonies in space. As we acquire a better and better knowledge of the planets, man will set out once again on a new kind of expansion, as millions of years ago he set out to invade our own planet from his tiny starting point in East Africa.

And in that case, what will happen to our space colonizers?

If they remain for a long time on another planet, isolated from their earthly habitat, they will change, shift; both their biology and their culture will evolve differently. Just imagine all the new cultures that could derive from living on other planets, and perhaps whole new species as well.

If we go out into space, our bodies will change considerably, isn't that true? Even the astronauts who have spent any time orbiting the Earth prove that the human organism doesn't function up there the way it does on Earth, that the bones, for instance, tend to atrophy rapidly. We risk becoming brainy slugs.

We still know relatively little about the conditions of life in space, and therefore about the consequences. When you're living and working under weightless conditions, there are important bodily changes that take place. For one thing, the mineral elements of the bones migrate, and it's difficult to get them back to their original sites. After several million years in space, our cousins will doubtless be quite different from ourselves. If man populates a number of planets, it's entirely possible that there will be great diversity in each of the different planetary populations, even different races.

Diversity: today we're in the process of losing it throughout the world. The world is becoming smaller; there's a globalization taking place, as well as a homogenization of peoples.

That's quite true. People travel a lot more and a lot longer than they ever used to; they migrate culturally, intermarry. But when we see various ethnic tribes, such as the Native Americans, being relegated to what are euphemistically called "reservations," we have to ask ourselves a very important question: if the purpose is to allow them to preserve their traditions, their songs, their language, isn't it also preventing them from having normal access to the contemporary world? Are these reservations, these little self-contained native islets, more for our pleasure than for theirs? I strongly believe that these native populations have no other solution than to mingle genetically and culturally with the rest of the world—and we with them—or disappear. Nostalgia has its place, but not if it isolates people.

In your opinion, is the movement toward complexity that's been taking place ever since the Big Bang still going on?

Yes. The accumulation of knowledge is forever expanding. Man is progressing toward greater knowledge, more freedom, a culture and perhaps a nature that are more and more complex. We are following the same path as did matter and life itself.

So can we label you an optimist?

Unreservedly. I find that human societies are increasingly well organized. We're becoming more and more aware of our precious environment, and even though we're often slow to act, the awareness is there. The United Nations, despite all its difficulties, is truly a forum for understanding. When you look at the world with even a bit of perspective, you can see that man's awareness of the world condition is far greater than it was only seventy years ago. And what is seventy years measured in the context of human history?

In the context of human history, nothing. In the context of human life, a great deal.

We also have to remember that the duration of modern life is negligible if we compare it with the three million years of our species' existence. Humanity today, although it has indeed attained a certain level of reflective thought, still strikes me as relatively young. A great many problems we've experienced in the present century are a result of various peoples' not having access to sufficient information.

The Future of the Universe

The life span of a human being is ridiculously small compared with that of all human history. We have just pointed this out with Yves Coppens. Is it possible that we are still in the prehistoric period of the universe? And if so, how long will it go on?

HR: The most recent observations seem to favor the theory of continuing expansion of the universe. Under this scenario, the universe is thought to be infinite in dimension, and therefore its life would be infinite. It continues to cool off, as it slowly tends toward a temperature of absolute zero. This said, one cannot be definitive about the subject. Our predictions are based on the existence of four forces, and four only. There is absolutely no reason for us to assume that others will not be discovered in the future. These discoveries, if and when they come, could modify our predictions.

If the universe is infinitely expanding, this means that it is becoming more and more empty as time goes by, that the heavenly bodies will go on moving away from one another. That, in turn, means that the sky, seen from Earth, will at some point become completely black.

The stars that shine in our night sky are not part of that expansion. Globally speaking, they are not moving away from us. The expansion is taking place between the galaxies, not within them. With the passage of time, these galaxies will indeed appear less and less visible to our telescopes. But it will be several billion years before this diminution occurs significantly.

All that is hypothetical, since man won't be around to make the observations. Some stars are going to die, notably ours, the Sun. Isn't that true?

Yes. As we've said earlier, at this point our Sun has already burned half its hydrogen. It's a middle-aged star. In five billion years it will have used up almost all its hydrogen, at which point it will become a red giant. Its core will contract more and more, whereas its atmosphere will expand until it reaches about a billion kilometers across. At the same time, its color will change from yellow to red.

And at that point, all the planets in the solar system will be burned up, incinerated.

Yes. The Sun will be a thousand times brighter than it is today. Seen from Earth, it will occupy a large part of the sky. The temperature of our planet will rise to several thousand degrees. Life will disappear, the Earth will be volatilized. That will take several hundred million years. Our star will disintegrate, as will Mercury, Venus, and perhaps Mars. The planets farther out from the Sun, such as Jupiter and Saturn, will lose their hydrogen and helium atmospheres; all that will be left will be their enormous, bare, rocky cores. Still later, the Sun, stripped of its nuclear energy, will look like a white dwarf, about the size of the Moon. Over a period of several billion years, it will slowly grow cold and become a dark dwarf, a lightless stellar corpse.

What will become of the matter Earth was made of?

It will return to interstellar space, where, later on, it may serve as the basis for new stars, even take part in the formation of planets.

And new lives?

Why not? Perhaps one day in the future, in some far-off biosphere, other living organisms will be made up of atoms from our bodies.

The only thing we know for sure is that man cannot remain on Earth for more than another four billion years.

Yes, but we can speculate, and even think of it as a distinct possibility, that, as Professor Coppens has suggested, long before that fateful date we will be able to go on long interstellar voyages. Just think of the progress that has been made in two short generations: our grandparents traveled at most at thirty miles an hour, whereas today we have vessels capable of going thirty thousand miles an hour. It's not out of the question that we will invent space probes that will travel at almost the speed of light, in which case our descendants will be able to set off in search of light near distant stars.

I remember the lovely observation that Konstantine Tsiolkovski, the father of the Russian space program, made some years ago: "Earth is the cradle of man, but we don't remain forever in our cradle." All this said, the evolution of complexity can go on with man's involvement, but it can also continue without him. After all, there's a real chance that we aren't the heroes of this history.

You're absolutely right. We can easily conceive that the human species might well become extinct but that life on the planet would continue. Insects, for instance, can live with a higher degree of radioactivity than it takes to kill us. Insects could survive a nuclear war, develop intelligence, and go on to rediscover technology. And in several million years, assuming that they did, they would probably run into the same kind of pollution problems we have.

Up until now, we have carefully avoided trying to give our story a "meaning," or at least to adopt a determinist point of view. But everything we've seen leads to the assertion that we've continually moved toward increasing complexity. One can make a good case that it's going to continue.

I'm struck by two facets of reality. The first portrays this wonderful story of evolution that we've been describing, and there are all sorts of reasons to believe that the story has meaning. The second, darker facet reveals contemporary man incapable of living in harmony with his fellow man and with the biosphere. Wars and devastating setbacks are rife, as if at a certain point in the process of evolution something went awry.

And what conclusion do you draw from that?

One has to ask oneself: why are things going so well in the physical world and so poorly in the human? Is it possible that nature has attained its "level of incompetence" by progressing as far as it has in complexity? That, I imagine, might be one interpretation based solely on the effects of natural selection from a Darwinian viewpoint. But if the essential end product of evolution was that man should be free, is it perhaps the price we're paying for that freedom? We can sum up the cosmic drama in three phases: nature engenders complexity; complexity engenders efficiency; efficiency may destroy the complexity.

Which means?

In the twentieth century, human beings have invented two ways of autodestruction: nuclear weapons in sufficient quantities to destroy the planet, and the deterioration of the environment. The basic questions are: is complexity viable? Is it a good idea for nature to reach this level of evolution that leads it to threaten itself? Is intelligence a gift laced with poison?

And what's your answer?

*

At the present time, we're close to capacity on the planet Earth. Is it really possible for ten billion people to coexist without deterioration setting in? Even if humans are talented, which they have proved to be in thousands of ways, from breaking the atom to exploring space, that task—coexisting in these numbers—will be a greater challenge than any we have faced in the past. Among other things, it means for business that the notion of constantly expanded production has to go; it has to be replaced by the idea of "lasting developments." Understandably, that's hard for today's businesspeople to grasp, much less implement.

In other words, we have to manage the planet along the lines Joel de Rosnay suggests, or, as you see on bumper stickers, "Think locally, act globally."

In any organism there is a built-in alarm system that activates when the organism is in trouble. When one is wounded, the entire body is mobilized into action. We have to invent some sort of similar system for the planet. The United Nations and various worldwide humanitarian associations are an important first step. We need to take them much further.

Aren't we in some ways the victims of our myopia? Do we have our noses glued too closely to our own time? Or maybe we're simply still in a period of prehistory, as Professor Coppens suggests. Maybe we still have a long way to go to reach a higher stage of morality and civilization.

*

Has humanity really made progress in the realm of morality? I'm far from certain. To be sure, in the past couple of centuries, we've seen the abolition of slavery and the declaration of the rights of man. But Native Americans had already attained a remarkably high degree of human behavior hundreds of years ago. They had established rules of social conduct that influenced the drafters of the U.S. Constitution. Claude Lévi-Strauss also demonstrated convincingly that the very notion of slavery occurs with the appearance of major civilizations. No, whether we've made any true moral progress is open to question.

Isn't it possible that other civilizations, in other worlds, are facing the same problems we are?

The discovery of very robust forms of life around submarine geysers and in deep geological strata certainly gives weight to the hypothesis of extraterrestrial life. Our earthly civilization is in all likelihood only one among many. Going on the assumption that cosmic evolution has led to the formation of other planets, other forms of life, other intelligent beings, we can also presume that these extraterrestrial creatures have had to face the same kinds of threats that we face today on Earth. A visit to these worlds would in all likelihood reveal one of two very different faces: an arid planet, devoid of life, covered with the remains of radioactivity—in other words, a world where those who lived there failed to adapt; or green, welcoming surfaces on those worlds where the inhabitants did adapt.

*

Symbiosis or death, as Joel de Rosnay said. Or, put another way: wisdom or the revenge of matter.

The crucial question we face today is simply this: are we capable of coexisting with our own self-made power? If we are not, then evolution will go on without us. Like Sisyphus, we will have pushed our rock up to the mountaintop, only to have it escape our grasp and come crashing back down. What could be more stupid? We should not blind ourselves to the seriousness of the current situation. Despite all the potential negatives, however, my feeling is that we must remain optimistic. We have to do everything in our power to save the planet before it is too late. We're the party responsible, the inheritors. It's up to us to take whatever steps are required to make sure that this miracle of a planet of ours not only goes on but flourishes.